Seeing Clearly

The Impact of Atmospheric Turbulence on the Propagation of Extraterrestrial Radiation

S. Businger and T. Cherubini, Editors

"Seeing Clearly: The Impact of Atmospheric Turbulence on the Propagation of Extraterrestrial Radiation," edited by S. Businger and T. Cherubini. ISBN 978-1-60264-698-8.

Published 2011 by Virtualbookworm.com Publishing Inc., P.O. Box 9949, College Station, TX 77842, US. ©2011, S. Businger and T. Cherubini. All rights reserved. No part of this publication may be reproduced, stored in a retrieval system, or transmitted in any form or by any means, electronic, mechanical, recording or otherwise, without the prior written permission of S. Businger and T. Cherubini.

Manufactured in the United States of America.

Table of Contents

Introduction

Introduction

There has been considerable progress in recent years to observe and quantify the effects of atmospheric turbulence and refractivity gradients on telescope image stability at various wavelengths—collectively referred to as seeing. At the same time there has been progress in modeling these atmospheric effects to support advances in methods that mitigate resulting telescope image degradation. Given the multidisciplinary nature of these efforts, there is substantial opportunity for further advances. This textbook on seeing assembles contributions from experts in the various facets of the complex challenge posed by atmospheric turbulence on optical and longer wavelengths. It is our hope that this introductory text will inspire students to explore this exciting area.

The first part of Seeing Clearly provides an introduction to atmospheric turbulence from the perspective of the meteorologist and the astronomer. Turbulence, with its eddies reaching into nearly every discipline of fluid dynamics, has proven to be a treacherous subject for scientists, stranding even specialists in its whirls of nonlinearities and statistical conundrums. Most flows in nature are turbulent. In the Earth's atmosphere jet streams in the upper troposphere display turbulence; cumulus clouds are in turbulent motion; and the Earth's boundary layer, where the effects of the surface are felt, is turbulent. The primary motivation of meteorologists to study turbulence in the atmosphere is to obtain tractable expressions for the fluxes of heat, momentum, water vapor, and other atmospheric constituents in the boundary layer between the Earth's surface and the free atmosphere.

Atmospheric turbulence is a primary concern in ground-based astronomy because it dramatically impacts the angular resolution of a telescope. Small temperature and moisture fluctuations in the atmosphere result in variations of the atmospheric refractive index. The wave front of radiation traveling through the atmosphere changes as it encounters inhomogeneities in the refractive index, degrading optical image quality. Turbulent fluctuations of the atmospheric refractive index are described by the refractive index structure function, C_N^2. The maximum telescope resolution is defined by a parameter called seeing, which is proportional to the integral of C_N^2. Seeing is expressed in units of arc seconds and describes the angle occupied by the star image at the full-width and half-maximum of its intensity profile. The C_N^2 and seeing are commonly used by astronomers to describe the turbulent state of the atmosphere at the time of their observations.

In Part 2 the varied instrumentation developed for observing optical turbulence and the principles under which they operate are described. Instability of atmospheric flows is related to the interaction of viscous terms and nonlinear inertia terms in the equations of motion. The dimensionless Reynolds number gives a measure of the ratio of the inertial forces to the viscous forces. Turbulence occurs at high Reynolds numbers and is dominated by inertial forces, which tend to produce random eddies, vortices, and other flow instabilities. In this case, because the equations of motion are nonlinear, each individual flow pattern has certain unique characteristics that are associated with its initial and boundary conditions. No general solution to the Navier-Stokes equations is known; consequently, no general solutions to the problems in turbulent flow are available. This situation means that empirical data are vital to understanding turbulent effects in a given environment. One of the challenges for astronomers and meteorologists alike is understanding how to reconcile the often-diverging observations produced by different instrument approaches.

Although the nature of turbulence is familiar from watching the formation of eddies in smoke rising from a cigarette or a smoke stack, it is very difficult to give a precise definition. In this regard it is useful in this introduction to briefly describe the characteristics of turbulence. Observations show that turbulence is random, chaotic in nature, making a deterministic approach to turbulence problems impossible; instead, one has to rely on empirical and statistical methods. Turbulence is diffusive, causing rapid mixing and increased rates of momentum, heat and mass transfer. Turbulence is rotational and three-dimensional; high levels of fluctuating vorticity characterize turbulence. For this reason, vorticity dynamics play an essential role in the description of turbulent flows. Turbulent flows are always dissipative. Viscous shear stresses perform deformation work that increases the internal energy of the fluid at the expense of the kinetic energy of the turbulence. Thus, turbulence needs a continuous supply of energy. Turbulence is a continuum phenomenon, governed by the equations of fluid mechanics. Even the smallest scales occurring in a turbulent flow are ordinarily far larger than any molecular length scale. Finally, turbulence is not a feature of fluids, but of fluid flows. The characteristics of turbulence depend on its environment. Therefore students of turbulence concentrate on families of flows with fairly simple boundary conditions, like boundary layers, jets, and wakes.

In part 3 methods that mitigate telescope image degradation resulting from atmospheric turbulence are described. Adaptive optics (AO) is a technology used to improve the performance of optical systems by reducing the effect of wavefront distortions. It is used in astronomical telescopes and laser communication systems to remove the effects of atmospheric distortion. Adaptive optics works by measuring the distortions in a wavefront and compensating for them with a deformable mirror or a liquid crystal array, which modulates the spatial phase of the incoming light. Horace Babcock first envisioned AO in 1953, but it did not come into common usage until advances in computer technology during the 1990s made the technique practical. There are a number of physical limitations to AO performance, leading to successive generations of more and more sophisticated techniques.

Knowledge of the distribution and magnitude of atmospheric turbulence can be used by astronomers to help optimize AO operation, providing a motive for modelers to predict atmospheric turbulence. Part 4 describes the challenges of modeling atmospheric optical turbulence and recent progress in developing an operational system to provide seeing forecasts. Numerical modeling of optical atmospheric turbulence holds great promise for assisting astronomy. In addition to optimizing the performance of adaptive optics, model output can i) provide 2- and 3-dimensional maps of the atmospheric turbulence over extended areas to serve as guidance for forecasters and telescope operators, ii) serve as guidance for telescope queue scheduling, because different turbulence conditions favor different types of instrument observations, and iii) be archived to form a simulated climatology of atmospheric turbulence and seeing for use in site monitoring and selection.

A great challenge in modeling optical turbulence comes from the fact that it occurs on spatial scales much smaller than regional weather models can resolve. Optical turbulence is a sub-grid scale phenomenon that must be parameterized in the model in a way that is informed by empirical observations of the turbulence. This circumstance poses a limitation in the ability of the model to predict the full spatial and temporal variability of optical turbulence. As model resolution improves and computational capabilities increase, the ability to reproduce the range of variability of the turbulence will also improve.

Numerical models represent an important complementary tool to instruments that measure optical turbulence and together they provide a better understanding of the nature of atmospheric optical turbulence. Finally, human forecast experience in combining observations, model guidance, and detailed knowledge of the mountain-weather environment is the third component needed in the operational effort to produce accurate forecasts of seeing.

This introductory textbook on seeing focuses on the observational, modeling, and mitigation aspects of the problem of atmospheric optical turbulence on astronomy. It brings together expertise in both meteorology and astronomy. It is the authors' hope that this book will give students and young scientists the background to inspire new instrument and modeling approaches.

For the reader's clarity the parameters which appear in the book and that are commonly used to define the optical turbulent state of the atmosphere are defined in the short preface that follows.

The editors would like to express our deep appreciation for the hard work of the technical editor, May Izumi, and the graphic artist, Nancy Hulbirt without whom this book would not be what it is. The editors are also indebted to the many reviewers who provided thoughtful input regarding early drafts of the chapters in this book. The preparation and publication of this book was made possible through support from the Nation Science Foundation under awards ATM-0644453 and AGS-0946581.

Steven Businger and Tiziana Cherubini, Editors

Preface
Definitions of some useful parameters

Throughout the book the authors of the various chapters will refer to parameters that are commonly used to define the optical turbulent state of the atmosphere. For the reader's convenience we list their definitions here.

The atmospheric turbulence strength is measured by the refractive index structure constant $C_n^2(\vec{x},t)$. A detailed explanation and definition of C_n^2 will be provided in Chapter 2. In astronomical applications it is usually assumed that turbulence is stratified horizontally and therefore $C_n^2(z,t)$ indicates the turbulence profile. The typical unit for the turbulent profile is m$^{-2/3}$. A number of useful parameters can be derived from the turbulence profile. Their definitions follow below and they are summarized in Table 1.

- The *Fried number* or coherence length r_0 (Roddier, 1981) represents the equivalent diameter of a telescope whose angular resolution is not strongly limited by the atmospheric optical turbulence.
- *Seeing* ε_0, defined as the width at the half height of a star image at the focus of a large diameter telescope (Roddier, 1981), describes the image quality/resolution of a telescope.
- The *isoplanatic angle* θ_0 represents the maximum angular separation of two stellar objects producing similar wavefronts at the telescope entrance pupil (Roddier et al., 1982; Fried, 1979)
- The *wavefront coherence time* τ_0 (Roddier, 1982): for turbulence moving with a horizontal velocity V, τ_0 defines the time the wavefront remains coherent (temporal isoplanatism). The wavefront coherence time τ_0 fixes the minimum exposure time for a detection system in order to obtain a non-aliased signal and can be linked to the bandwidth necessary for the adaptive optic system (see Chapter 6) to keep up with the temporal variation of the wavefront (Travouillon et al., 2009).

These parameters are sometimes referred to as astro-climatic parameters by astronomers and depend on the telescope-observing wavelength λ. Throughout this book, unless otherwise indicated, $\lambda=0.5$ μm is chosen as a representative wavelength for optical astronomy.

Table 1. Relevant astro-climatic parameters; L indicates the optical path, λ the wavelength at which the telescope is observing, V the magnitude of the horizontal wind velocity.

Parameter	Definition	Typical Units
Generic moment of the turbulence profile	$\mu_n = \left[\int_0^L z^n C_n^2(z)dz \right]^{-3/5}$	$[m^{1/3+n}]$
Fried Number	$r_0 = \left[0.423 \left(\dfrac{2\pi}{\lambda} \right)^2 \int_0^L C_n^2(z)dz \right]^{-3/5}$	$[m]$
Seeing	$\varepsilon = 0.98 \dfrac{\lambda}{r_0}$	$[arcsec]$
Isoplanatic Angle	$\vartheta_0 = \left[2.91 \left(\dfrac{2\pi}{\lambda} \right)^2 \int_0^L z^{5/3} C_n^2(z)dz \right]^{-3/5}$	$[arcsec]$
Coherence Time	$\tau_0 = \left[2.91 \left(\dfrac{2\pi}{\lambda} \right)^2 \int_0^L V^{5/3} C_n^2(z)dz \right]^{-3/5}$	$[ms]$

Part I
Atmospheric Turbulence

Atmospheric Turbulence from the Perspective of a Meteorologist

The planetary boundary layer (PBL) is a tropospheric layer close to the earth's surface. It has a depth that exhibits strong diurnal variation, particularly over land. This height is typically about 1000 m in the day time and of the order of 100 m during nights. This significant diurnal variation in the height of the PBL is caused by the variations in the surface energy budget whose components include incident short wave solar radiation on clear days and outgoing long wave radiation from the earth's surface during nights. Clear sky conditions with no clouds produce most significant diurnal variations. The PBL is also the layer whose processes are governed by turbulence, which in turn depends on atmospheric stability determined by the vertical temperature gradients. The energy budget at the surface has different components that include net radiation, turbulent sensible and latent heat fluxes, and ground heat flux. The surface energy budget influences the variations in the ground temperature and also the air temperature near the surface. Although horizontal mean advection is also a factor in the magnitude of air temperature, vertical surface turbulent heat flux is a major factor. Turbulence is largest near the surface decreasing to near zero values at the top of the PBL.

Another aspect is the large concentrations of aerosols in the PBL due to the existence of multitude of sources near the surface. Similar to the variation in turbulence, aerosols also sharply decrease to small concentrations at the top of the PBL, particularly during the day time hours. An exception occurs where distinct land plumes form above the local planetary boundary layer caused by advection of aerosols and gases from a distant source. They originate in a land convective boundary layer (typically about a kilometer depth) and are transported over the locally formed boundary layer such as the marine boundary layer (Simpson and Raman, 2004).

Planetary boundary layer exists over all surfaces irrespective of topography. In other words, they do exist over hills and mountain peaks although their structure and characteristics are greatly influenced by the topographical effects. These can include drainage flows at night as well as differential heating of mountain slopes and the associated local circulations during the day time.

As mentioned before, boundary layer turbulence is very much affected by atmospheric stability. Convective conditions exist during daytime with air temperatures decreasing with height sharply in the surface layer. This is a layer close to the surface, typically about 100 m during day time and smaller in the nights. Above the surface layer, turbulence associated with large eddies mix the PBL air thoroughly causing the PBL structure to be well mixed throughout. This is reflected in all the mean variables such as air temperature, wind speed, moisture, and aerosols

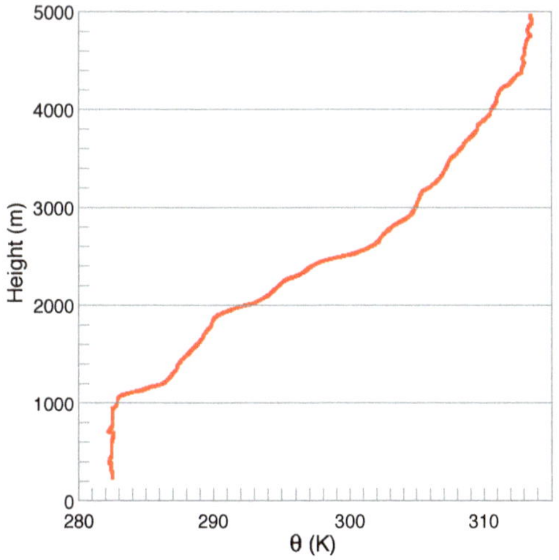

Figure 1.1   Variation of potential temperature (theta) as a function of height at 13:00 local time over a land site. Planetary Boundary Layer (PBL) is characterized by strong gradient in air temperature near the surface, and a mixed layer capped by a temperature inversion. Height to the base of the inversion is generally taken to be the PBL height.

being near constant in this "mixed layer" as shown in Fig. 1.1 for the air temperature over land. This mixed layer is capped by an elevated inversion which essentially typifies an increase in air temperature with height. This statically stable layer aloft acts much like a lid on pollution generated in the boundary layer. Water vapor is also well mixed in the planetary boundary layer, but decreases sharply at the top of the layer (Fig. 1.2).

Another mitigating factor that influences PBL structure and turbulence is the coastal circulation caused by diurnal variations in horizontal surface heat fluxes (or surface temperatures) between the ocean and land. These variations cause sea breezes during the day time and land breezes in the nights. Winds at the shore line become onshore with the onset of sea breezes and offshore with the land breezes. A thermal internal boundary layer forms near the coast due to sharp discontinuity of the surface roughness and temperature. The effect of land–water interaction on the PBL and the lower troposphere can be felt at large distances as the manifestation of a land plume laden with dust and gas originating form a distant polluted land boundary layer. This plume can affect visibility drastically at a location far removed from the source.

The objective of this chapter is to present and discuss the observations that characterize the diurnal variation of boundary layer turbulence. As a contrasting example, we will look at how the turbulence varies between a cloudy day and a clear day.

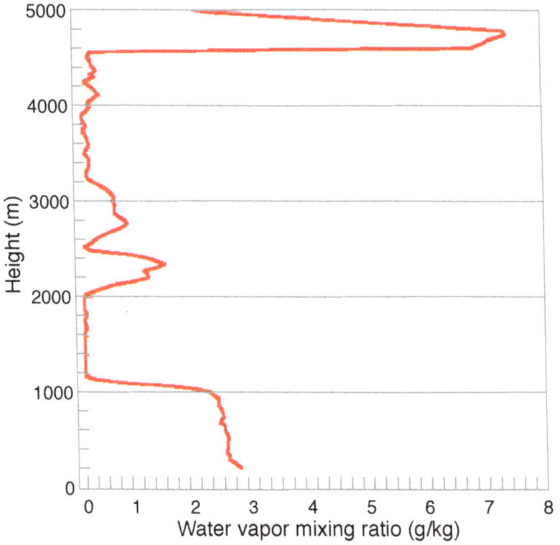

Figure 1.2 Variation of water vapor mixing ratio or moisture as a function of height.

1.1 Statistical Description of Turbulence

Turbulence being random is best defined using statistics. Turbulence is characterized as fluctuations from the mean value of the variable. Thus the fluctuations can be positive or negative depending on the magnitude of the instantaneous value. The fluctuations essentially manifest turbulent eddy characteristics. The time period to average the observations to obtain the mean value should take into account most of the energy containing eddies. This is about one hour for the boundary layer. Statistical parameters are often used to characterize turbulence. They include variance and turbulent kinetic energy (TKE). The latter is half the sum of the variances in three directions. In Eq. 1.1 instantaneous wind velocities in three directions, U, V, and W are split into mean and fluctuating parts,

$$U_j = \overline{U}_j + u'_j \tag{1.1}$$

where subscript "j" represents components in three different directions (U, V, and W) and the bar denotes an averaged value. Similarly instantaneous air temperature θ can be split into mean temperature $\overline{\theta}$ and fluctuating temperature θ'.

Covariances $\overline{u'_i u'_j}$ and $\overline{u'_i \theta'}$ obtained through the averaged product of two fluctuating quantities provide a measure of fluxes. When the covariance of vertical velocity fluctuations and temperature fluctuations, $\overline{w'\theta'}$ is considered, it becomes a turbulent flux. We are most interested in vertical turbulent fluxes since maximum gradients in wind speed, air temperature, and moisture occur in this direction. Vertical turbulent fluxes also cause changes in the mean variables in the boundary layer.

Turbulent Kinetic Energy typifies turbulence in a broad sense since it is the sum of the three variances. It is a measure of turbulence directly related to momentum, heat and moisture transport in the PBL. Budget of the TKE provides an insight into how the turbulence is generated, transported and dissipated. Following is a TKE ($\bar{e}$) budget equation:

In Eq. 1.2 term 1 is time rate of change of TKE often referred to as inertial or storage, term 2 is advection by mean wind, term 3 is buoyancy production (this can also be a loss of TKE depending on its sign), term 4 is shear production (always positive), term 5 is turbulence transport, term 6 is pressure transport, and term 7 is dissipation. By estimating different terms and comparing, one can understand the relative importance of various competing processes on the generation and maintenance of turbulence.

$$\underbrace{\frac{\partial \bar{e}}{\partial t}}_{1} + \underbrace{\bar{U}_j \frac{\partial \bar{e}}{\partial x_j}}_{2} = \underbrace{\delta_{i3} \frac{g \overline{(u_i' \theta')}}{\bar{\theta}}}_{3} - \underbrace{\overline{u_i' u_j'} \frac{\partial \bar{U}_i}{\partial x_j}}_{4} - \underbrace{\frac{\partial \overline{u_j' e}}{\partial x_j}}_{5} \qquad (1.2)$$

$$\underbrace{- \frac{1}{\bar{\rho}} \frac{\partial \overline{(u_i' p')}}{\partial x_i}}_{6} - \underbrace{\varepsilon}_{7}$$

1.2 Observing Turbulence

Best way to study turbulence is to perform an observational analysis that brings out its characteristics. We have performed an analysis of turbulence observations measured at location, Cyril, Oklahoma. A regional map showing this location is given in Fig. 1.3. Cyril is in a region that is geographically horizontally homogeneous. This makes it a good place to study boundary layer turbulence. Because of horizontal homogeneity, one can neglect horizontal gradients of surface variables such as heat fluxes. Data used in this study includes measurements of air temperature, incoming solar radiation, wind speed, atmospheric turbulence, momentum flux, and sensible heat flux.

The Surface Meteorological Observation System (SMOS) is used by the ARM program (Atmospheric Radiation Measurement) program (www.arm.gov) to observe atmospheric mean conditions on the ground. This analysis will use 30 minute measurements of surface wind speed and air temperature. Wind speed is measured at a height of 10 m using a wind propeller anemometer. Air temperature is measured using Vaisala T-RH probe at a height of 2 m.

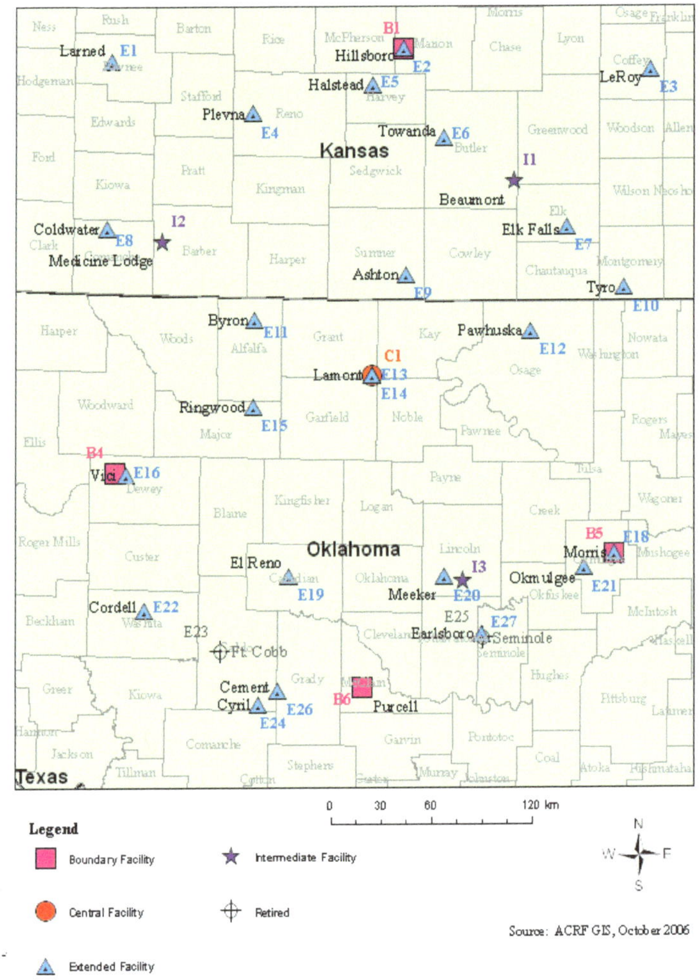

Figure 1.3 A regional map showing the locations of various stations where mean and turbulence parameters were measured. Location of Cyril, Oklahoma is highlighted on the southern portion of the map.

Data from a Solar Infrared Radiation Station (SIRS), also part of the ARM program, was used to obtain measurements of incoming solar radiation in the form of shortwave broadband direct normal irradiance. These measurements were obtained in 15-minute intervals using a pyrheliometer at 2 m above the ground.

The Eddy Correlation Flux Measurement System (ECOR) is used to observe atmospheric surface layer turbulence data. A fast response three dimensional sonic anemometer is used in measuring high frequency variations (about 20 Hz) of the three components of winds, temperature, and humidity. Fluctuating variables are then deduced after subtracting the mean values obtained over a period of one hour. The velocity and temperature fluctuations are then correlated with each other to obtain variances and covariance. Covariance involving u, v, and w components provide momentum flux. Covariance involving vertical velocity (w) and temperature fluctuations provides an estimate of heat flux and the one involving humidity and vertical velocity fluctuations provide the water vapor flux or the latent heat flux.

1.3 Solar Radiation and Boundary Layer Turbulence

Incoming solar radiation (Fig. 1.4) during the day time is the main driving force for the generation of PBL turbulence. One way to examine the influence of the solar radiation is to analyze observations on two consecutive days, one with clouds and one without. In fact, one of the best approaches to study this problem is when a total solar eclipse occurs during mid day and the solar radiation is completely cut off for a few minutes. Such studies have been conducted before (SethuRaman, 1982; Raman et al., 1990). Results indicated a total change in local atmospheric stability even though solar radiation is slowly diminished during a total solar eclipse between the first and the second contacts. Turbulence in the PBL also decreased significantly at this time thus demonstrating the dominant influence of the incoming solar radiation on the boundary layer structure and stability. However, occurrence of total solar eclipses during mid day over land without weather related clouds are rare. Next best approach to study the role of the solar radiation on turbulence characteristics is to compare the observations between a cloudy day and a cloud free day. The advantage of analyzing for turbulence with cloud as factor is that one can examine diurnal variations. The disadvantage is the likelihood of several layers of clouds and the duration of cloudiness.

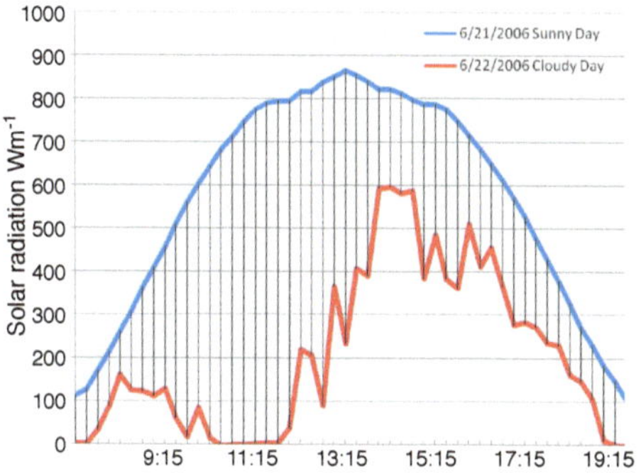

Figure 1.4   A comparison of incoming solar radiation for June 21 (sunny day) and June 22, 2006 (cloudy day) during the daylight hours.

Data from June 21 and 22, 2006 were analyzed to examine the effect of the presence of clouds on turbulence. During both days, Cyril (Fig. 1.3) remained in a warm, humid, pre-frontal air mass. The weather was mostly sunny throughout the first day according to satellite imagery (not shown). However, throughout the second day, a weather front remained nearly stationary to the north of Cyril and provided mostly cloudy conditions. Also during the second day, a brief period of rain occurred early morning and there were likely a few breaks in the clouds during the afternoon. The goal of this analysis is to examine the variations of the mean air temperature, mean wind speed, turbulence and the production and dissipation terms in the TKE budget on these two consecutive days, one sunny and one cloudy to demonstrate the influence of solar radiation on the surface layer turbulence.

1.3.1 Solar Radiation

Since incoming shore wave solar radiation is the main driving force for the PBL during day time, a comparison of the solar radiation between a cloud free day (June 21) and a cloudy day (June 22) is shown in Fig. 1.4. Values of observed short wave solar radiation for both the days have been plotted. Observations for June 21 (clear day) show a parabolic variation from near zero at the sun rise (about 0700 LT) to a maximum value of 850 W m^{-2} at 1300 LT, then decreasing to zero at sunset. On the contrary, on June 22, when clouds were present, solar radiation was diminished to a great extent in the morning. However, the values increased to about 500 Wm^{-2} by late afternoon as the skies became partly cloudy. Being a summer day, total cloudiness for the entire day was not present. It will be interesting to see how the PBL turbulence varied for these two days with different incoming short wave solar radiation.

1.3.2 Mean Air Temperature

Hourly mean air temperature measured at a height of 10 m for both the days are presented in Fig. 1.5 from 0600 LT on June 21 to 0600 LT on June 22. As expected, the mean air temperature (Fig. 1.5) during the sunny day (June 21) is warmer and has a larger diurnal range of about 13 K as compared to only 3 K on the cloudy day (June 22). The broader diurnal range is due to the presence of larger solar radiation for a longer period of time. Because Cyril remained in the warm sector of the weather front for both days, synoptic conditions remained similar. Thus, most of this difference can be attributed to the cloud cover and its interception of radiation. Incoming short wave radiation is intercepted by clouds during the

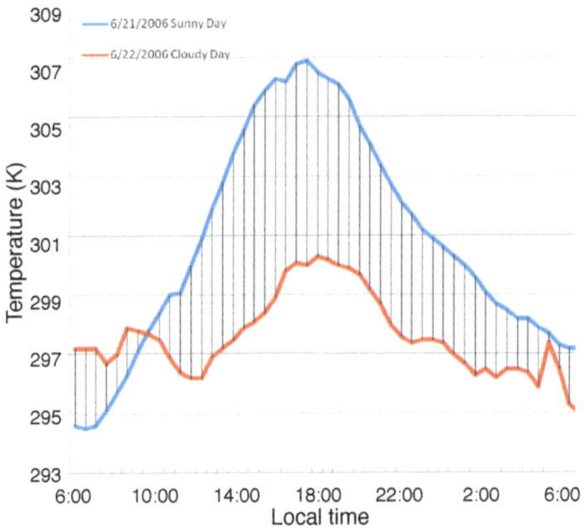

Figure 1.5 Diurnal variation of 2 m air temperature on June 21 and June 22, 2006.

day, thus causing cooler air temperatures. Outgoing long wave radiation is intercepted at night by clouds preventing the ground surface from cooling excessively thus causing warmer air temperatures as observed at 0600 LT on June 22. Air temperature at this time is about 3 K higher than similar observations on June 21, contributing to a decrease in the diurnal range of temperature values.

1.3.3 Mean Wind Speed

The synoptic weather pattern is often the biggest factor affecting the mean wind speed. When the synoptic pattern is not a factor as was the case for the analyzed sunny day (June 21), the mean wind speed near the surface is often influenced by the atmospheric stability and has a diurnal pattern somewhat similar to that of mean air temperature. Wind speed near the surface is low just before sunrise due to the existence of stable atmospheric conditions with the air temperature increasing with height. Immediately after the sunrise, as the incoming solar radiation heats up the surface, PBL becomes unstable or convective resulting in increased turbulence and vertical mixing. This process brings down higher momentum air resulting in an increase in mean wind speed. This process continues until about noon after which decrease in the solar radiation has the opposite effect. Variation of mean wind speeds averaged over a period of 30 minutes for June 21 (cloud free day) and June 22 (cloudy day) are shown in Fig. 1.6. The mean wind speed during the sunny day increases throughout the morning hours, peaks in the late afternoon, and then decreases sharply in the evening. This occurs as incoming radiation heats the surface, causing the PBL to deepen thus allowing for more efficient mixing of higher momentum air from aloft. This vertical momentum transfer begins as soon as the surface begins to heat up and continues through the morning and early afternoon before weakening with the loss of heating. During the

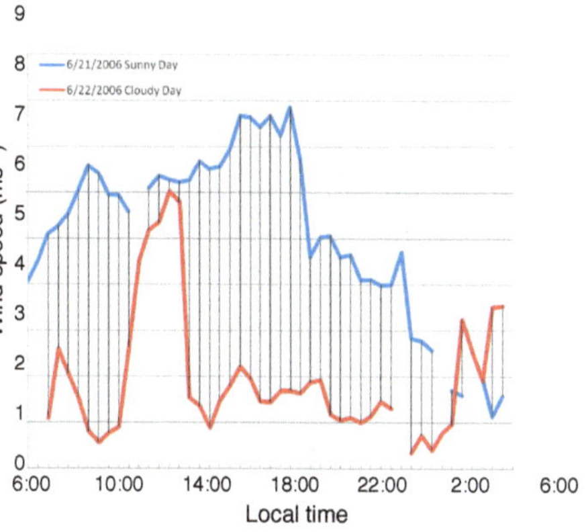

Figure 1.6 Variation of hourly averaged wind speed on June 21 and June 22, 2006.

cloudy day (June 22), a peak in the wind speed around 12:00 LT is associated with the low level convergence caused by a local storm. The remainder of the day is characterized by light winds. No diurnal pattern is evident on June 22 because of the lack of surface heating due to the cloud cover. Variation in wind speed in turn affects the variation in turbulence through the shear production term.

1.3.4 Turbulent Sensible Heat Flux

The (turbulent) sensible heat flux is best explained as the amount of energy from the sun that actually gets converted into heat at the surface. Near surface sensible heat flux is estimated from the measurements of covariance between vertical velocity fluctuations and temperature fluctuation. The sensible heat flux pattern for the sunny day (June 21) shows a bell shaped curve with minimum values before sunrise and maximum in the early afternoon hours. Sensible heat flux is negative during the night time indicating small downward turbulent heat flux. This is caused by stable conditions of the atmosphere with air temperature increasing with height resulting in downward heat flux.

During the day time the sensible heat flux becomes positive or upward as the incoming solar radiation heats up the surface and reached a maximum magnitude of about 550 Wm^{-2}.

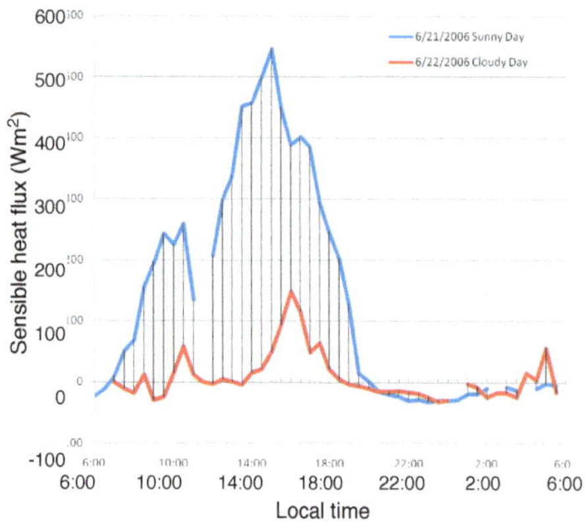

The extensive cloud cover on the second day blocks a significant amount of the incoming short wave radiation, decreases the surface

Figure 1.7 Diurnal variation of the turbulent sensible heat flux on June 21 and June 22, 2006.

temperature, and decreases the turbulent sensible heat flux near the surface. Sensible heat fluxes on the cloudy day are several orders of magnitude less than that on the cloud free day. The brief positive spike in sensible heat flux on the second day during the early afternoon is likely the result of a break in the clouds. A brief negative value in the sensible heat flux during early morning is probably the result of a brief local storm. This time corresponds to rain showers on radar observations at this location. Rain showers moisten the soil which then cools through evaporation. Advection of relatively warmer air over the cool soil then produces a negative or downward sensible heat flux. After sunset sensible heat flux values are negative (or downward) indicating clear nights on both days.

1.3.5 Velocity Variances

The variance of velocity is a measure of the intensity of turbulence and is strongly related to the sensible heat flux. Variance of vertical velocity fluctuations are shown in Fig. 1.8 for June 21 and June 22. When the sensible heat flux is the highest, i.e., during the middle of the sunny day, the vertical variance is the

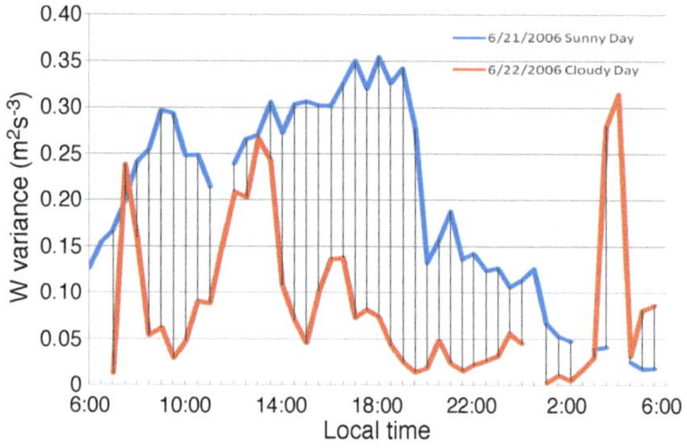

Figure 1.8 Diurnal variation of the vertical (w) variance component of turbulence on June 21 and June 22, 2006.

highest due to the effect of buoyant thermals. On June 22 (cloudy day), vertical turbulence is lower by a factor of about three except during the morning when a local storm was present. Velocity variance values approach zero during night time and the formation of a stable layer near the surface. Large values observed on June 23 early morning hours is believed to be due to the breaking of internal gravity waves present during stable atmospheric conditions. During strong stable conditions, internal gravity waves coexist with turbulence even near the surface and break due to shear instability causing a sharp increase in turbulence (SethuRaman, 1977; SethuRaman et al., 1982). For both days, horizontal components of velocity variance contributed far more to turbulent energy than the vertical component (not shown).

1.3.6 Turbulent Kinetic Energy

Turbulent Kinetic Energy (TKE) which is half of the sum of the three velocity variances is a good indicator of turbulence in the PBL since it is the sum of the three variances. It is a parameter often used in modeling the PBL to represent turbulence. TKE again depends on atmospheric stability. TKE values for a sunny day (June 21) and a cloudy day (June 22) are compared in Fig. 1.9. The TKE has a diurnal variation with larger values during day time when both shear and buoyancy productions of turbulence are present. The TKE on June 21 reaches a maximum value of 3.5 $m^2 s^{-2}$ during late afternoon and corresponding values for the cloudy day are much smaller (about 0.75 $m^2 s^{-2}$). For a time period in the morning, the TKE values on both days are similar possibly caused by the local storm activity discussed above. A large spike in the TKE values during the early morning hours of June 23 is believed to have been caused by the breaking of internal gravity waves discussed above. Now we will examine the production and dissipation terms of turbulence for the two days.

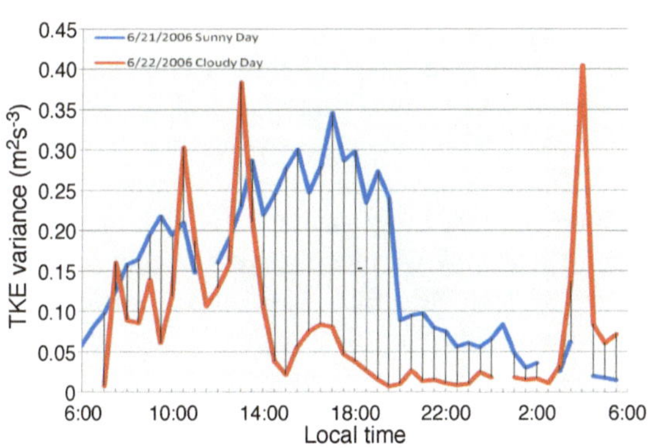

Figure 1.9 Diurnal variation of turbulent kinetic energy (TKE) on June 21 and June 22, 2006.

1.3.7 Buoyancy Production of TKE

As indicated by turbulence kinetic energy budget (Eq. 1.1), turbulence is produced by two processes, shear and buoyancy. Shear production is positive all day since friction causes wind speed to become zero at the surface. However, buoyancy production has a diurnal variation over land. It is positive during day time due to the heating of the surface by the incoming solar radiation and negative during nights due to the cooling of the surface by outgoing long wave radiation. Buoyancy production of turbulence for the clear day (June 21) and the cloudy day (June 22) are shown in Fig. 1.10. It is the largest on the sunny day and during late afternoon consistent with the variation of sensible heat flux (Fig. 1.7). Values during the nighttime hours are negative thus decreasing the turbulence generated by wind shear as seen in the variation of TKE (Fig. 1.9).

On the cloudy day, buoyancy production is significantly less (by a factor of about four) during day time. Negative values of buoyancy production just after 0800 LT on the cloudy day are likely the result of rain showers discussed before. The buoyancy production stays generally very low throughout the rest of the cloudy day besides one spike in the afternoon probably associated with some partial clearing. A comparison of the variation of the buoyancy term (Fig. 1.10) and the sensible heat flux (Fig. 1.7) reveals that their shape is almost identical for both days, indicating the dependency of buoyancy production of turbulence on sensible heat flux and consequently on cloud cover.

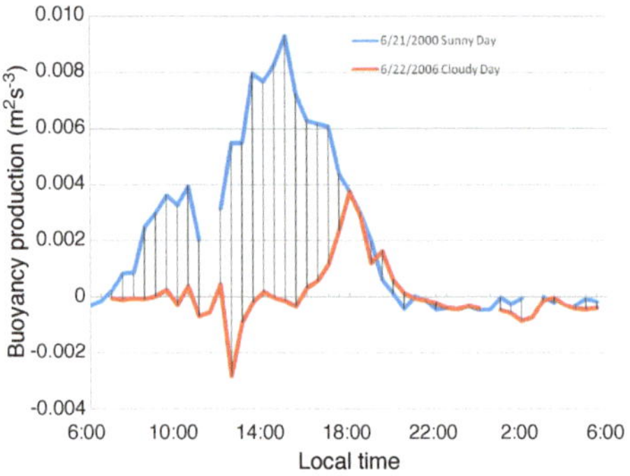

Figure 1.10 Diurnal variation of the buoyancy production of the TKE on June 21 and June 22, 2006.

1.3.8 Shear Production of TKE

Vertical variation in wind speed always exists due to the presence of friction at the surface. This vertical variation causes wind shear which in turn generates turbulence as indicated in the TKE budget (Eq. 1). In fact, the shear production of turbulence is the process through which turbulence derives its energy from the mean flow and is a production term in the TKE budget during day and night. Values of shear production estimated using the velocity covariance and mean wind shear for June 21 and June 22 are shown in Fig. 1.11. Variation of shear production is similar to the variation of mean wind speed shown in Fig. 1.6 for both days. Larger wind speeds lead to larger wind shear and more shear production.

Values of shear production on the sunny day (June 21) are much larger than the cloudy day by factors varying from four to seven except during the morning of June 22 when a local storm was present. Shear production on June 21 (sunny day) displays a bell curve over the course of the day with turbulence production increasing in the morning before decreasing rapidly immediately after sunset. As would be expected this variation is comparable (and directly related) to the variation of variances and

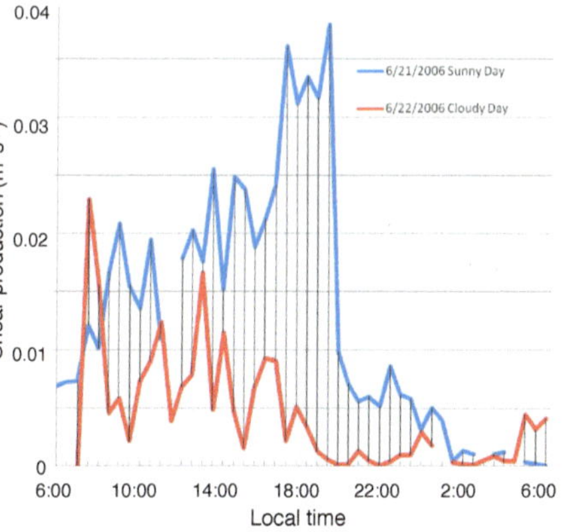

Figure 1.11 Diurnal variation of the shear production of the TKE on June 21 and June 22, 2006.

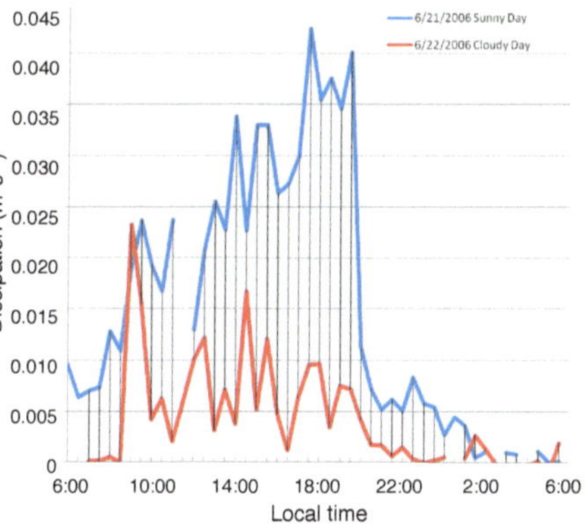

Figure 1.12 Diurnal variation of the dissipation of TKE on June 21 and June 22, 2006 estimated from the budget.

TKE discussed earlier. The shear production on the cloudy day is generally much lower. This is to be expected due to lower wind speeds observed on the cloudy day. Shear production gets close to zero during the evenings of both the days as winds become calm.

1.3.9 Dissipation of TKE

Turbulence is eventually dissipated through internal friction caused by the viscosity of air and is always a loss term in the TKE budget equation. The dissipation of TKE is estimated using the TKE budget equation (Eq. 1.1) neglecting the two transport terms. Turbulent transport and pressure redistribution (or pressure transport) term can be neglected close to the surface as a first approximation. Variation of dissipation essentially follows the variation of TKE; higher the turbulence, larger the dissipation. Thus larger values of dissipation on the first day are attributed to the larger amount of TKE available within the PBL to dissipate. Dissipation is higher during day time when TKE is higher, while TKE dissipation gets close to zero during nights. Therefore, on a cloudy day dissipation values are smaller in line with the TKE values.

1.4 Land Plume and Turbulence

A major factor in affecting visibility in remote places and islands is the presence of a land plume about a kilometer deep above the PBL. It is a manifestation of turbulence in the PBL from a distant source. An example of such a land plume over the Arabian sea is shown in Fig. 1.13.

Observations were taken from an aircraft using a downward looking lidar (Simpson and Raman, 2004). Aerosol concentrations in this plume located at an altitude of 2 km are remarkably high even after the plume has traveled a distance of about 1000

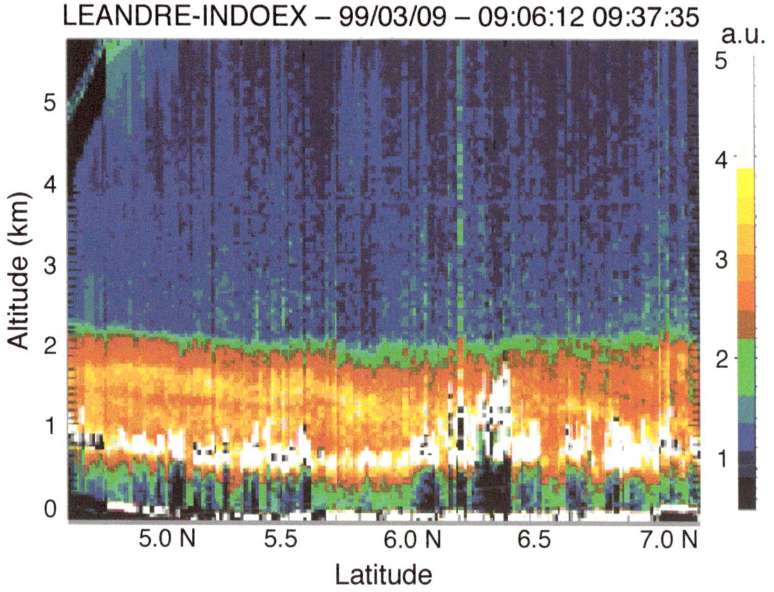

Figure 1.13 Elevated layer of aerosols over the Arabian Sea observed during INDOEX (1999) with a downward looking lidar mounted on an aircraft. This observation was taken at a downwind distance of about 1000 km from where the plume originated.

km. Role of turbulence in this land plume is the daytime mixing over land (or an urban area) to a height of about 2 km. There is substantially lower turbulence over the ocean near the coast caused by the advection of warm air over relatively cooler water and the formation of a stable air mass. As the aerosol laden air mass travels offshore, it becomes an elevated, distinctly different layer as shown in Fig. 1.13. This layer has a decaying turbulence, but still well mixed. On top of this layer, the stable layer persists mainly due to large scale subsidence (Fig. 1.14). At large distances from the coastline, the marine boundary layer becomes convective as the near surface air adjusts to the sea surface temperature below. During INDOEX (Indian Ocean Experiment) conducted during 1999, it was found that the land plume extends for distances exceeding 1000 km. It eventually gets mixed down by the strong updrafts present near the Intra Tropical Convergence Zone (ITCZ).

1.5 Summary

Turbulence in the Planetary Boundary Layer is mainly driven by the solar radiation. Wind shear and thermal buoyancy are the two production mechanisms. Because of the variations in these productions, there is a strong diurnal variation in turbulence. Shear production of turbulence is related to wind speed gradient in the lowest layers of the atmosphere and is affected by large scale weather

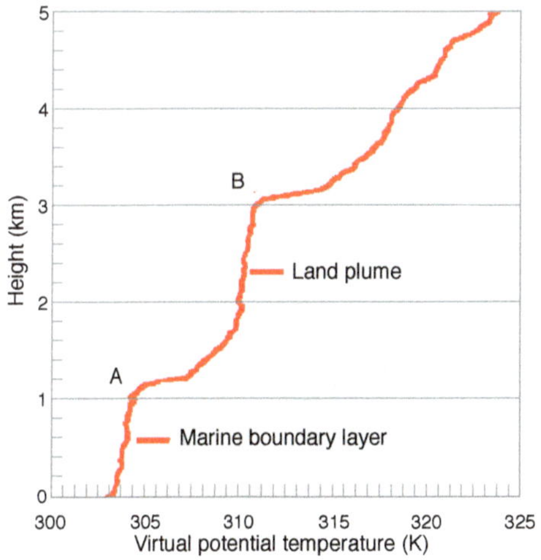

Figure 1.14 Vertical temperature profile over the ocean showing the marine boundary layer and the elevated residual (mixed) layer where the plume exists.

processes. Buoyancy on the other hand is related to surface—air temperature gradient which has a diurnal variation. During the day time, earth's surface is heated by the incoming solar radiation resulting in an upward turbulent sensible heat flux and positive buoyancy. With both production terms being positive, turbulence tends to be larger during day time. During nights, on the other hand, the surface cools due to outgoing long wave radiation resulting in negative buoyancy. This leads to suppression of any turbulence generated by the wind shear. An example showing the contrast between a cloudy day and a clear day presented in this chapter shows these processes well.

Acknowledgements

Support for this study was provided by the Division of Atmospheric Sciences, National Science Foundation under Grants ATM-0233780 and ATM-0342691.

Authored by
S. Raman, M. Diaz, and S. Lanham
Department of Marine, Earth, and Atmospheric Sciences,
North Carolina State University, Raleigh, NC

CHAPTER 2

Optical Turbulence for Astronomy

2.1 Classical Kolmogorov Theory

The spatio-temporal behavior of the turbulence in our atmosphere is described within the classical theory of Kolmogorov (1941). Refractive index fluctuations are closely related to temperature fluctuations through the Gladstone formula: temperature fluctuations can only occur (1) in a medium with temperature gradient and (2) with dynamical turbulent motion. Dynamical turbulent motions within a homogeneous thermal field will not produce any refractive index turbulence, also called "optical turbulence". Nor will a strong temperature stratification within a laminar medium produce any optical turbulence.

The classical description of dynamical turbulence, the formation of turbulent eddies, was given by Kolmogorov (1941). Motion is described by a three-dimensional vector but temperature is a scalar parameter. Temperature, like other scalars such as ozone concentration and humidity, is considered as a tracer or additive. Additives might be passive (if they do not interact with the motion) or conservative (if they do not change their scalar value during the motion). The spatio-temporal fluctuations of additives described by Obukhov (1949) and Yaglom (1949) are discussed in Section 2.1.3. As refractive index is related to temperature, its spatio-temporal fluctuations follow the same law that is described in Section 2.1.4. But we will demonstrate that temperature is not passive or conservative, which implies a new vision as explained in Section 2.2. We will show a paradigmatic example of optical turbulence appearing within a pair of "laminae" in Section 2.2.1. Our hypothesis is reinforced by a numerical simulation made by Werne and Fritts (1999).

2.1.1 Atmospheric Stability

In order to understand turbulent motion in our atmosphere, it is important to have a better insight to its stability. For what concerns astronomy, two regimes are easily related to day or to night time. During the day, the sun illuminates the ground, which transfers heat to the atmosphere. The air at ground level is hotter than the air on top, giving rise to convection. Even without any wind, turbulent cells are growing within the convective layer, one to two kilometers thick. During the night, when the sky is clear, the ground surface radiatively loses its heat and becomes colder than the air above, giving rise to a stable stratification. Why?

Let us define a temperature that remains independent of its altitude, when the pressure decreases exponentially. This is the so-called potential temperature that remains constant in our atmosphere for a moving parcel of air without any heat exchange. Equation of state of air, assuming an ideal gas, follows the Clapeyron law:

$$p = \rho R T \qquad (2.1)$$

where p is the pressure in Pa, ρ the density in kg m^{-3}, T the absolute temperature in K, and $R = 287$ J K^{-1}kg^{-1}, a constant for dry air. During an adiabatic process for an ideal gas, one can write:

$$\frac{dT}{T} = \frac{\gamma - 1}{\gamma} \frac{dp}{p} \qquad (2.2)$$

$$\gamma = \frac{C_p}{C_v} = \frac{7}{5} \qquad (2.3)$$

$$\frac{\gamma - 1}{\gamma} = \frac{2}{7} = 0.286 \qquad (2.4)$$

where C_p and C_v are respectively the specific heat for constant pressure and the specific heat for constant volume.

For a static fluid, one knows that:

$$dp = -\rho g \, dz \qquad (2.5)$$

where we assume that z grows with altitude, g is the gravitational acceleration and dp an incremental motion. From Equations 2.5 and 2.2, one finds that:

$$\frac{dT}{dz} = \frac{\gamma - 1}{\gamma} \frac{g}{R} = -\gamma_a = -0.0098 \text{ K m}^{-1} \qquad (2.6)$$

i.e., almost 1 degree per 100 m. Temperature decreases linearly with altitude

$$T = T_0 - \gamma_a (z - z_0). \qquad (2.7)$$

Now, one can define the potential temperature as the temperature of a parcel of air adiabatically brought to the 1000 hPa pressure level. We integrate Eq. 2.2 as follows:

$$\int_{T_0}^{T_1} \frac{dT}{T} = \frac{\gamma - 1}{\gamma} \int_{p_0}^{p_1} \frac{dp}{p} \qquad (2.8)$$

which leads to

$$Log \frac{T_0}{T_1} = \frac{\gamma - 1}{\gamma} Log \frac{p_0}{p_1} \tag{2.9}$$

and finally to:

$$T_1 = T_0 (\frac{p_0}{p_1})^{\frac{\gamma - 1}{\gamma}} \tag{2.10}$$

Thus, the potential temperature $\theta = T_1$ of a parcel that was at temperature T_0 and pressure level p_0 and has been brought to pressure level $p_1 = 1000$ hPa:

$$\theta = T (\frac{1000}{p})^{\frac{2}{7}}. \tag{2.11}$$

Now, one can define the static stability of a parcel of air, depending upon the sign of $d\theta/dz$, as one can see in Fig. 2.1.

Three cases of stability: neutral, unstable, and stable, depending upon the sign of $d\theta/dz$

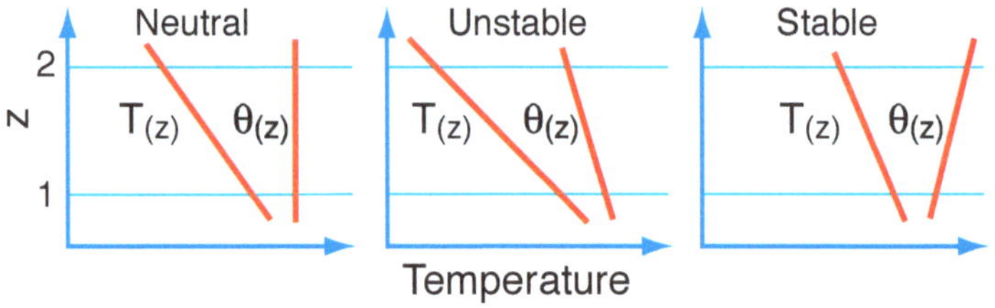

Figure 2.1 Three cases of stability: neutral, unstable, and stable, depending on the sign of $d\theta/dz$.

- Neutral: $d\theta/dz = 0$, which means that the absolute temperature decreases adiabatically (Eq. 2.2). In this case, a parcel of air moved from its original position will remain at its new position.
- Unstable: $d\theta/dz < 0$, which means that the absolute temperature decreases more rapidly than adiabatically. In this case, a parcel of air, since its density is less than the density on top, will move vertically, like a hot bubble within a cold medium. Turbulent motion will start without any need of kinetic energy. It is a turbulent convective layer.
- Stable: $d\theta/dz > 0$, which means that the absolute temperature decreases less than adiabatically. In this case, a parcel of air, since its density is more than the density above, will remain at the same altitude.

2.1.2 Kolmogorov Classical Theory for Motion

2.1.2.1 Condition for turbulent flow apparition
- **Turbulence in homogeneous medium, Reynolds number.** If one considers a flow with a characteristic speed U, a characteristic dimension L, and a viscosity $v[\text{m}^2\text{s}^{-1}]$, the kinetic energy per unit volume is

$$E_k = \rho U^2 \tag{2.12}$$

and the energy dissipated through viscosity forces is

$$E_v = \rho \frac{vU}{L^2} L = \rho \frac{vU}{L}. \tag{2.13}$$

In order to compare the ratio between kinetic energy and viscosity energy, Reynolds (1883) introduced the dimensionless Re number:

$$Re = \frac{E_k}{E_v} = \frac{UL}{v}. \tag{2.14}$$

It is acknowledged that a flow becomes unstable and turbulent when Re is larger than its critical value, around 2000.

- **Turbulence in a stratified medium.** In the previous section we considered only inertial and viscosity forces. In our atmosphere, other forces such as buoyancy and Coriolis might be taken into account. As we are only interested in dimensions from few meters to few hundred meters, Coriolis forces can be neglected. At this scale, viscosity forces can also be neglected when compared to potential forces generated by the density stratification. Potential energy is due to a $\Delta\rho$ density variation across a vertical variation Δz. Force F applied to a unit volume of air with a density difference $\Delta\rho$ when compared to ambient medium is $F = g\Delta\rho$. The potential energy is the work of F over Δz:

$$E_{pot} = g\Delta\rho\Delta z. \tag{2.15}$$

For what concerns kinetic energy, it depends upon variation of speed ΔU over the same height Δz:

$$E_k = \rho \Delta U^2. \tag{2.16}$$

Richardson introduced the ratio between E_{pot} and E_k, which is a dimensionless number:

$$Ri = \frac{E_{pot}}{E_k} = \frac{g\Delta\rho\Delta z}{\rho\Delta U^2} \tag{2.17}$$

which can also be written:

$$Ri = \frac{\frac{g\Delta\rho}{\rho\Delta z}}{\left(\frac{\Delta U}{\Delta z}\right)}. \tag{2.18}$$

In Fig. 2.2 we summarize the various flow situations depending upon the Richardson number and its critical value 1/4.

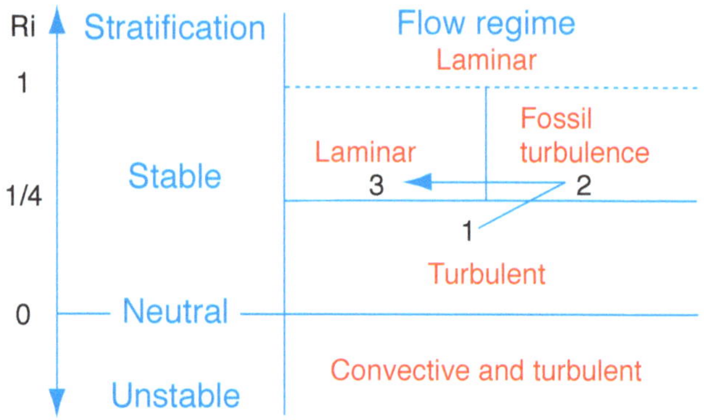

Figure 2.2 Type of stratification and flow regime depending upon the Richardson number. When Richardson number drops below its critical value 1/4, the laminar flow becomes turbulent (1). Then, likely after some kinetic energy dissipation, the Richardson number increases to about 1 (2) and the flow remains turbulent (Fossil turbulence), and later it comes back to laminar (3).

2.1.2.2 Kolmogorov cascade

In 1941, Kolmogorov presented his "cascade" description of large turbulent eddies breaking into smaller and smaller eddies until they disappear when their kinetic energy is dissipated in heat by viscosity forces. Fig. 2.3 sketches this process (The reader can find a good description by Frish [1995].).

Let us analyze this cascade at scale l, the eddies have a time scale τ_l given by:

$$\tau_l = \frac{l}{u_l} \tag{2.19}$$

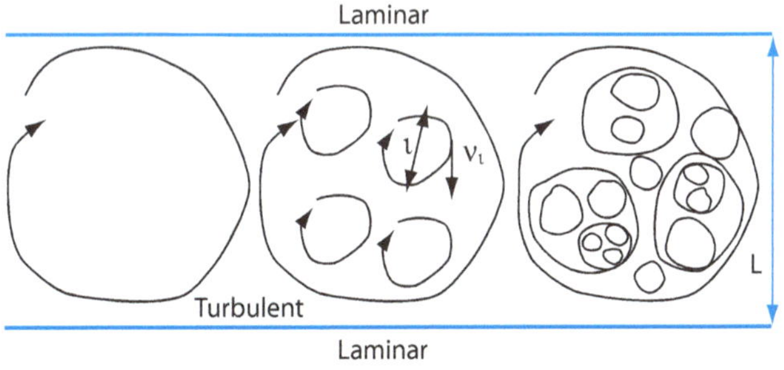

Figure 2.3 Kolmogorov's cascade: within a turbulent layer of thickness L, large eddies are developing which yield part of their kinetic energy to smaller and smaller eddies.

where u_l is the characteristic velocity of this eddy. The kinetic energy which is transmitted to a smaller eddy, per mass unit is u^2_l . The rate of kinetic energy transmitted from eddies of size l, per mass unit and per time unit is

$$\varepsilon_l = \frac{\delta u^2}{\delta t} = \frac{u^2_l}{\tau_l} = \frac{u^3_l}{l} . \qquad (2.20)$$

2.1.2.3 Structure function of the velocity

The velocity structure function is defined through

$$D_u(l) = <(u(x) - u(x+l))^2> \sim u^2_l \qquad (2.21)$$

which is a function of scale l. Kolmogorov assumed that there is a range of scales l, $l_v < l < L$, where viscosity can be neglected. More, he assumes that the structure function $D_u(l)$ depends only upon l and ε:

$$D_u(l) = F(\varepsilon, l) \qquad (2.22)$$

and that the rate of energy transfer is independent of the eddy size. Equation 2.22 might be written as follows:

$$D_u(l) = C\varepsilon^\propto l^\beta \qquad (2.23)$$

where C is a dimensionless constant. From Eq. 2.21, we establish its dimension:

$$Dim[D_u(l)] = L^2 T^{-2} \qquad (2.24)$$

where L and T are respectively the dimensions of length and time. From Eq. 2.20 one can find the dimension of ε:

$$Dim[\varepsilon] = Dim\left[\frac{u_l^2}{\tau_l}\right] = Dim\left[\frac{u_l^3}{l}\right] = L^2T^{-3}. \tag{2.25}$$

From Eqs. 23, 24 and 25, one can write:

$$L^2T^{-2} = L^{2\alpha}T^{-3\alpha}L^{\beta}. \tag{2.26}$$

We solve the linear system:

$$2\alpha + \beta = 2 \tag{2.27}$$
$$\alpha = 2/3 \Rightarrow \beta = 2/3 \tag{2.28}$$

and thus obtain the canonical form of the structure function of velocity variations:

$$D_u(l) = C\varepsilon^{2/3}l^{2/3}. \tag{2.29}$$

In Eq. 2.29, $\varepsilon^{2/3}$ corresponds to the "intensity" of the turbulence while $l^{2/3}$ is the spatial dependency. One must keep in mind that this law is valid within the viscous dissipation scale and the "external" scale L:

$$l_v < l < L. \tag{2.30}$$

2.1.2.4 Viscous dissipation scale l_v

One can imagine that kinetic and viscous dissipation are almost similar at the viscous dissipation scale, meaning that the condition expressed in the definition of the Reynolds number (Eq. 2.14) might satisfy the following relationship:

$$Re = \frac{UL}{v} \sim 1. \tag{2.31}$$

From Eq. 2.20, the viscous dissipation rate ε_v is:

$$\varepsilon_v = \frac{u_v^3}{l_v}. \tag{2.32}$$

From Eq. 2.31 we bring u_v into Eq. 2.32 to find l_v:

$$l_v = (v^3/\varepsilon)^{1/4}. \tag{2.33}$$

One notes that the viscous dissipation range becomes smaller when turbulence (ε) is more intense.

2.1.2.5 Characteristic velocity and time at scale l

From Eq. 2.20, we can establish:

$$u_l = (\varepsilon l)^{1/3} \tag{2.34}$$

$$\tau_l = \frac{l}{u_l} = \varepsilon^{-1/3} l^{2/3}. \tag{2.35}$$

τ_l is the time of a turnover.

2.1.2.6 Graphical representation of the velocity structure function

In Fig. 2.4 is plotted the trend of the velocity structure function. The range between l_v and L is the so called "inertial range" because motions are dominated by inertial forces, and viscous forces are negligible. On the other hand, no eddy is larger than L, and hence the structure function saturates.

To summarize, kinetic energy is injected in the medium by the largest eddies of size L. It is admitted that L is from a few hundreds of meters to one km (Gavrilov et al., 2005). On the other hand, the dissipation length l_v can be computed from Eq. 2.33. Let us imagine a turbulent flow with ΔU variation of 5 m/s over a thickness $L = 500$ m, $\varepsilon = \Delta U^3 / L = 0.25$, and knowing that the kinematic viscosity of the air is about 17.10^{-6} m²/s, one finds $l_v = 0.37$ mm.

2.1.2.7 Spectral density of the velocity

The "2/3" spatial dependency has the same counterpart in the frequency domain. Here, we will show that this law can also be inferred from dimension considerations. Correlation functions as well as structure functions can be deduced from spectrum density through Fourier transform, in both directions.

The one-dimensional correlation function of an isotropic and homogeneous field, say "f", $B_f(l)$, is related to the one-dimensional spectral density $V_f(K)$ through Tatarski (1961):

$$B_f(l) = \int_{-\infty}^{+\infty} \cos(Kl) V_f(K) dK \tag{2.36}$$

and the variances of the "f" fluctuations are:

$$< f^2 >= B_f(0) = \int_{-\infty}^{+\infty} V_f(K) dK. \tag{2.37}$$

The total kinetic energy satisfies:

$$< u^2 >= \int_{-\infty}^{+\infty} V_u(K) dK. \tag{2.38}$$

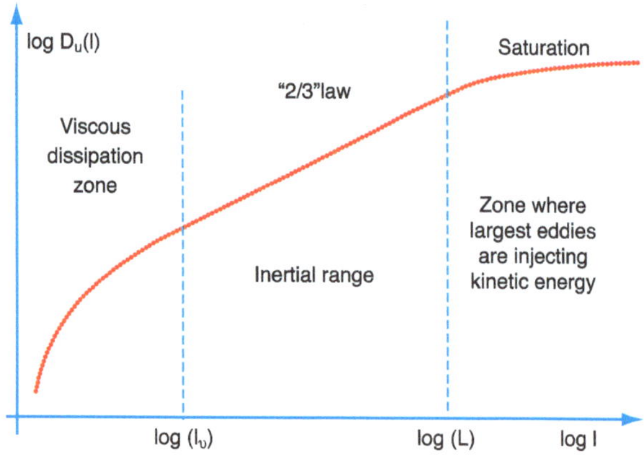

Figure 2.4 Sketch of the velocity structure function in log-log axis.

The fraction of kinetic energy which is passed through the increment of wavenumber δK can be deduced from Eq. 2.38:

$$\delta u^2 = V_u(K)\delta K \tag{2.39}$$

and now the unidimensional spectrum is written:

$$V_u(K) = \frac{\delta u^2}{\delta K}. \tag{2.40}$$

Kolmogorov hypothesis says that $V_u(K)$ depends only upon the rate ε of kinetic energy per mass unit and time unit, all along the cascade of K, and, as stated in Section 2.1.2.2, ε does not depend upon K and thus is constant all along the cascade:

$$V_u(K) = C'\varepsilon^\alpha K^\beta. \tag{2.41}$$

From Eq. 2.40, one can compute the dimension of Eq. 2.41:

$$Dim[V_u(K)] = \frac{L^2T^{-2}}{L^{-1}} = L^3T^{-2}. \tag{2.42}$$

From the dimension of ε expressed in Eq. 2.25, the dimension of Eq. 2.42 yields:

$$L^3T^{-2} = (L^2T^{-3})^\alpha(L^{-1})^\beta \tag{2.43}$$

from which one can write a linear system of equations:

$$\begin{cases} 2\alpha - \beta = 3 \\ 3\alpha = 2 \end{cases} \Rightarrow \begin{cases} \alpha = 2/3 \\ \beta = -5/3 \end{cases} \qquad (2.44)$$

which allows us to write the solution:

$$V_u(K) = C'\varepsilon^{2/3}K^{-5/3} \qquad (2.45)$$

which is the famous "−5/3" law of the one dimension spectrum of the velocity field. As stated by Tatarski (1961), if one considers a structure function of the form:

$$D_f(l) \propto l^p \qquad (0 < p < 2) \qquad (2.46)$$

then the one-dimensional spectral density might be:

$$V_f(K) \propto K^{-(p+1)}. \qquad (2.47)$$

This last equation is purely based on Fourier transform properties. Now, in our case, where $p = 2/3$, one finds $K^{-5/3}$ which is exactly the solution that we found in Eq. 2.45. It seems like magic!

In Fig. 2.5 one can see the behavior of the energy spectrum of velocity fluctuations as a function of the wavenumber $log\ K$. All along the cascade $\varepsilon_v = \varepsilon = \varepsilon_L$. Kinetic energy is injected at $K = 2\pi/L$ with a rate ε, and then cascades until it reaches the dissipation at $K = 2\pi/l_v$. Within the so-called "inertial range", the turbulence is assumed to be fully developed and isotropic.

2.1.2.8 Inertial range extension and Reynolds number

Let us estimate the range where Kolmogorov's law is applicable. For so doing, let rise Reynolds number to power 3, using 2.14:

$$Re^3 = U^3L^3/v^3 \qquad (2.48)$$

and using Eqs. 33 and 20 ($\varepsilon = U^3/L$):

$$Re^3 = \frac{L\varepsilon L^3}{L^4\varepsilon} = \left(\frac{L}{l}\right)^4 \qquad (2.49)$$

and thus the extension of the inertial range becomes:

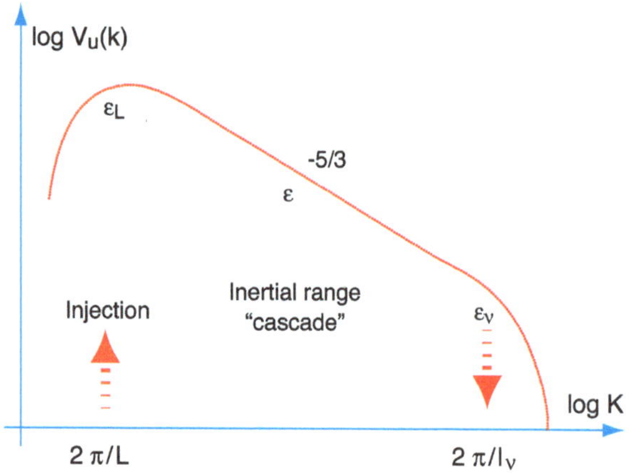

Figure 2.5 Sketch of the energy spectrum of velocity fluctuations in a log-log plot.

$$\frac{L}{l} = Re^{3/4}. \tag{2.50}$$

This indicates that the inertial range increases when turbulence becomes stronger, like the Reynolds number to power 3/4. It is easy to show also that the ratio between characteristic times increases with the square root of Reynolds number:

$$\frac{\tau_L}{\tau_l} = Re^{1/2} \tag{2.51}$$

and that characteristic velocities also increase:

$$\frac{u_L}{u_l} = Re^{1/4}. \tag{2.52}$$

In the above-mentioned situation (see Section 2.1.2.6) the Reynolds number is $Re = 1.5\ 10^8$ and

- $Re^{3/4} = 1.4\ 10^6$, $L = 500$ m and $l = 0.37$ mm
- $Re^{1/2} = 12\ 10^3$, $\tau_L = L/U = 100$ s and $\tau_l = 8$ ms
- $Re^{1/4} = 111$, $U_L = 5$ m/s and $u_l = 4$ cm/s

As one can see $u_L/u_l = Re^{1/4} \ll \tau_L/\tau_l = Re^{1/2} \ll L/l = Re^{3/4}$.

2.1.3 Obukhov and Yaglom Classical Theory for Passive Additive

Spatial and spectral properties of the velocity field have been established in the previous Section 2.1.1. But what are the properties of scalar fields like temperature, concentration of H_2O, O_3, CO_2, etc. It is clear that, for such a scalar component, turbulent fluctuations can occur only if:

$$\overrightarrow{grad}\ \theta \neq 0. \tag{2.53}$$

where θ is a spatial variation of this component. Let us imagine a flow where temperature is constant, even if turbulent; motion will mix parcels of the same temperature, and no θ fluctuation is expected. Now, suppose a laminar flow, but with a T_1/T_2 interface, as is sketched in Fig. 2.6. When ΔU increases, oscillations appear, as shown in Fig. 2.7. Then the Richardson number drops below 1/4,

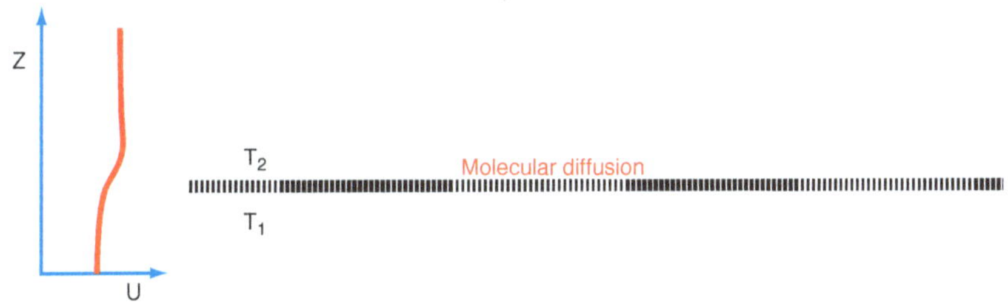

Figure 2.6 In a laminar flow, at the interface of two layers at T_1 and T_2, only molecular diffusion occurs.

Figure 2.7 At the beginning, left, the flow is laminar, then becomes oscillating and finally turbulent (From Dimotakis, Lye and Papantoniou, 1981).

triggering turbulence. At this stage, turbulence appears isotropic and fully developed, but soon, due to Kolmogorov's cascade, the medium homogenizes with viscous process.

A scalar additive, like θ, might be conservative and passive. "Conservative" means that its value within a small volume does not change during the motion. A counter-example is a chemical component that reacts with the medium. "Passive" means that this additive does not interact with the turbulent motion.

One can imagine that temperature is not passive, since temperature and density are related and density plays a major role within a gravity field as expressed in

Eq. 2.5. Nevertheless, let us assume that temperature is a passive and conservative additive in the following. Then in Section 2.2 we will reveal our new vision. Indeed, temperature is not conservative since it decreases with altitude as shown in Eq. 2.7. The right additive that remains constant with altitude is the potential temperature θ, as seen in Eq. 2.11.

2.1.3.1 Structure function of a conservative and passive additive θ

Earlier, we defined ε, the rate of kinetic energy transmitted from an eddy to a smaller eddy per time unit. Let us define, by analogy with velocity, the rate $\bar{N}$ of θ variance fluctuations per time unit and for an eddy of size l:

$$\bar{N} = \frac{\theta_l^2}{\tau_l} = \frac{\theta_l^2 u_l}{l}.$$ (2.54)

As for the velocity, again we will assume that the rate of inhomogeneities of the additive θ is constant all along the "cascade" and thus it is independent of scale l. Obukhov (1949) and Yaglom (1949) assumed that the structure of θ is a function of $\bar{N}$, ε and l only. Thus one can write:

$$D_\theta(l) = <(\theta(l_1) - \theta(l_2))^2> = F(\bar{N}, \varepsilon, l).$$ (2.55)

But, $\bar{N}$ and ε being constant all along the cascade, they do not depend upon l and Eq. 2.55 can be written as follows:

$$D_\theta(l) = a^2 \bar{N}^\alpha \varepsilon^\beta l^\gamma$$ (2.56)

where a^2 is a positive constant without dimension. The dimension of $\bar{N}$ is given by Eq. 2.56:

$$Dim[\bar{N}] = \theta^2 T^{-1}.$$ (2.57)

Using Equations 2.25 and 2.57, and knowing that the dimension of $D_\theta(l)$ is θ^2, one can write:

$$\theta^2 = [\theta^2 T^{-1}]^\alpha [L^2 T^{-3}]^\beta L^\gamma$$ (2.58)

which gives the following equation system:

$$\begin{cases} \alpha = 1 \\ \beta = -1/3 \\ \gamma = 2/3 \end{cases}$$ (2.59)

which yields the equation of the θ structure function:

$$D_\theta(l) = \frac{a^2\bar{N}}{\varepsilon^{1/3}} L^{2/3} \qquad (2.60)$$

which stands only within the "inertial range". Like the velocity, the passive additive θ follows the same "2/3" spatial power law (see Eq. 2.29). The structure function is expressed as:

$$\begin{cases} D_\theta(l) = C_\theta^2\, l^{2/3} \\ C_\theta^2 = a^2\bar{N}/\varepsilon^{1/3} \\ l_\theta < l < L \end{cases} \qquad (2.61)$$

where l_θ is the smaller scale where θ fluctuations homogenizes with dissipation, which will be detailed in Section 2.1.3.4.

2.1.3.2 Spectral density of a conservative and passive additive θ

As seen earlier, the wavenumber counterpart of spatial domain of the one-dimensional spectral density can be written (Tatarski, 1961):

$$V_\theta(K) = \frac{\Gamma(5/3)}{2\pi} \sin\frac{\pi}{3} C_\theta^2\, K^{-5/3}. \qquad (2.62)$$

2.1.3.3 Determination of the flux of θ variance: $\bar{N}$

This determination needs a long demonstration which can be found in Tatarski (1961). Here, I will give a summary based on the equivalence between the diffusion coefficient D, which is active at scale lengths l_θ, and the turbulent diffusion K, which is active at scale length θ_L.

Let us define the additive θ as a sum of its mean value $\bar{\theta}$ and its fluctuating part θ'.

$$\theta = \bar{\theta} + \theta'. \qquad (2.63)$$

The above mentioned equivalence can be written as follows:

$$\bar{N} = K\,(\overrightarrow{grad(\bar{\theta})})^2 = D\,\overline{(\overrightarrow{grad(\theta')})^2}. \qquad (2.64)$$

This an important result saying that, under a stationary regime, the flux of variance $\bar{N}$ of additive θ, per time unit, is equal to both 1) the product of the turbulent diffusion coefficient and the square of the gradient of the mean potential temperature and 2) the product of the θ diffusion coefficient and the mean square of the gradient of θ fluctuations.

2.1.3.4 Inner scale of θ additive

Equations 2.54 and 2.20 determine the θ additive and dynamic characteristics of the motion at scale l and one can write:

$$\bar{N} = \theta_l'^2 u_l / l \text{ and } u_l = (\varepsilon l)^{1/3} \tag{2.65}$$

and thus:

$$\theta_l'^2 = \bar{N} l^{2/3} / \varepsilon^{1/3}. \tag{2.66}$$

One can note the analogy with Eq. 2.60 of the structure function of θ:

$$D_\theta(l) = a^2 \bar{N} / \varepsilon^{1/3}. \tag{2.67}$$

The scale l_D where fluctuations of θ will homogenize can be inferred from:

$$\bar{N} = D \overline{(\overrightarrow{grad}(\theta'))^2} = D \, \theta_{l_D}'^2 / l_D^2. \tag{2.68}$$

and finally:

$$\theta_{l_D}'^2 = \bar{N} / D \, l_D^2. \tag{2.69}$$

Equalizing Eqs. 69 and 66, one finds the inner scale:

$$l_D = (D^3 / \varepsilon)^{1/4} \tag{2.70}$$

which immediately compares with viscous dissipation length:

$$l_v = (v^3 / \varepsilon)^{1/4}. \tag{2.71}$$

The diffusion coefficient of temperature is $D_T = 19.10^6$ m²s⁻¹, which is similar to viscous dissipation $v = 17.10^6$ m²s⁻¹. In a general manner, the diffusion coefficient of many additives are of the same order of magnitude.

2.1.3.5 External scale of a passive additive

In Eq. 2.61, we saw that:

$$C_\theta^2 = a^2 \bar{N} / \varepsilon^{1/3}. \tag{2.72}$$

In this formula, $\bar{N} = K (\overrightarrow{grad} \, (\bar{\theta}))^2$ is a "mean" parameter which depends upon the "mean" properties of θ, meanwhile $\varepsilon = u_l^3 / l$ is a "turbulent" parameter. As it was demonstrated in the case of additive θ, one can imagine the same kind of relationship for the velocity and write:

$$\varepsilon = K \, (\partial \overline{u}_i / \partial x_j)^2 \tag{2.73}$$

where $\overline{u}_i$ is the mean velocity field. The structure function of additive θ becomes:

$$C_\theta^2 = \frac{a^2 K (\overrightarrow{grad\,\theta})^2}{K^{1/3} \, (\partial \overline{u}_i / \partial x_j)^{2/3}} = a^2 \left[\frac{K^2}{(\partial \overline{u}_i / \partial x_j)^2} \right]^{1/3} (\overrightarrow{grad\,\theta})^2. \tag{2.74}$$

Now, let us scheme a simple stratified flow with a characteristic vertical thickness L as shown in Fig. 2.8. At a distance L, the structure function is:

$$D_\theta(L) = C_\theta^2 \, L^{2/3} = \Delta \theta^2. \tag{2.75}$$

But $\Delta \theta$ can be estimated from the gradient expression:

$$|\overrightarrow{grad\,\theta}| = \Delta \theta / L. \tag{2.76}$$

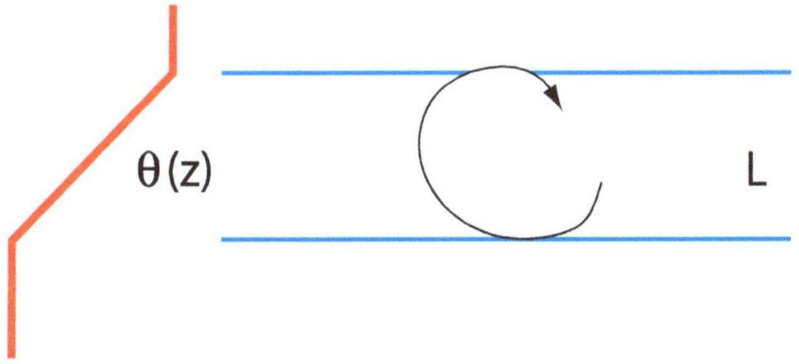

Figure 2.8 Scheme of a simple stratified turbulent flow of thickness L.

From Equations 2.75 and 2.76, one can infer C_θ^2 that we equalize with Equation 2.74 yielding:

$$C_\theta^2 = L^{4/3} \, (\overrightarrow{grad\,\theta})^2 = a^2 \left[\frac{K^2}{(\partial \overline{u}_i / \partial x_j)^2} \right]^{1/3} (\overrightarrow{grad\,\theta})^2 \tag{2.77}$$

from which we extract:

$$L^{4/3} = a^2 \left[\frac{K^2}{(\partial \overline{u}_i / \partial x_j)^2} \right]^{1/3} \tag{2.78}$$

and

$$L = a^{3/2} \left[\frac{K}{(\partial \bar{u}_i / \partial x_j)^2} \right]^{1/2} = a^{3/2} L_0. \tag{2.79}$$

We now define the external scale "L_0" which differs from L by a constant:

$$L_0 = \left[\frac{K}{(\partial \bar{V} / \partial x)} \right]^{1/2}. \tag{2.80}$$

One notes that the external scale depends only upon "mean" parameters: K and the vertical gradient of the mean horizontal wind. If we go back to Eq. 2.74, we can write:

$$C_\theta^2 = a^2 L_0^{4/3} \overrightarrow{(grad\,\bar{\theta})}^2. \tag{2.81}$$

This last equation allows us to estimate the turbulent parameter C_θ^2 from "mean" parameters which characterizes the mean flow such as the external scale L_0 and $\overrightarrow{grad\,\bar{\theta}}$.

2.1.4 Application to Refractive Index

As expressed in Section 2.1.1, because of exponential pressure decrease, the temperature decreases with altitude and thus it is not a conservative additive. That's why we introduced the potential temperature concept. In an adiabatic atmosphere:

$$d\bar{\theta}/dz = d\bar{T}/dz + \gamma_a = 0. \tag{2.82}$$

Using Eq. 2.81, one could compute the structure function of the temperature as:

$$C_T^2 = a^2 L_0^{4/3} \overrightarrow{(grad\,\bar{T})}^2. \tag{2.83}$$

In an adiabatic atmosphere where $\Delta\theta^2$ is zero, one would measure:

$$C_T^2 = a^2 L_0^{4/3} \gamma_a^2 \neq 0 \tag{2.84}$$

which is clearly false. Once again, the relevant parameter is the potential temperature, which can be expressed as:

$$C_\theta^2 = a^2 L_0^{4/3} \overrightarrow{(grad\,\bar{T} - \gamma_a)}^2. \tag{2.85}$$

and which is zero within an adiabatic medium.

Refractive index N is related to pressure p and temperature T through the Gladstone formula:

$$N = n - 1 = \frac{80.10^{-6}p}{T} = \alpha \frac{p}{T}.$$ (2.86)

The vertical gradient of the refractive index is thus:

$$\frac{dN}{dz} = \frac{\alpha}{T}\frac{dp}{dz} - \frac{\alpha p}{T^2}\frac{dT}{dz} = \frac{\alpha\rho g}{T} - \frac{\alpha p}{T^2}\left(\frac{d\theta}{dz} - \gamma_a\right).$$ (2.87)

In an adiabatic atmosphere where $d\theta/dz = 0$, one notes that $dN/dz \neq 0$. Hence, the refractive index N is not a conservative additive. That's why we introduce the potential refractive index N' which follows:

$$dN'/dz = -\alpha p/T^2$$ (2.88)

which is now conservative, and its structure constant is now:

$$C_{N'}^2 = \left(\frac{80.10^{-6}p}{T^2}\right)^2 C_\theta^2.$$ (2.89)

In the literature, despite all the precaution we took before, the reader will find C_T^2 instead of C_θ^2, and C_N^2 instead of $C_{N'}^2$. The refractive index structure function is written as:

$$\begin{cases} D_N(l) = C_N^2 l^{2/3} \\ l_N < l < L_0 \end{cases}$$ (2.90)

and the structure constant is:

$$\begin{cases} C_N^2 = a^2\,(K^2/\beta^2)M^2 \\ \quad = a^2 L_0^{4/3} M^2 \end{cases}$$ (2.91)

where

$$\begin{cases} M = dN'/dz = -80.10^{-6}p/T^2\,d\theta/dz \\ \beta = d\bar{U}(z)/dz. \end{cases}$$ (2.92)

2.2 New Vision

Within the astronomical community, high angular resolution and interferometry observations were interpreted through the main theoretical works of Tatarski (1961) and Roddier (1981). Both authors assumed the same spatial correlation or structure function deduced from Kolmogorov law within the inertial range $l_v < l < L$, bounded by the inner and outer scales. At this time, the outer scale was assumed to be the outer scale of motion, i.e. around few hundreds of meters, as stated by Gavrilov et al. (2005), and references therein.

Very few investigations were performed about the outer scale of temperature or refractive index. For example, Van Zandt (1978) established a stochastic model to relate mean vertical profiles of temperature and velocity and $C_T^2 (h)$ vertical profiles. In order to tune his model, he had to assume a constant outer scale of 10 m, which is much less than the dynamic outer scale.

To my knowledge, one of the first experimental proofs of a small outer scale of optical turbulence was performed by Mariotti and Di Bencdetto (1984), using an interferometer at various baselines. Only three baselines were explored, but he clearly demonstrated that for baseline $B >> 8$ m, the phase structure function is already saturating. Coulman et al. (1988) used Eq. 2.91 to estimate the $L_0(h)$ profile from the knowledge of $C_N^2 (h)$ and $M^2(h)$ profiles. They found that the optical turbulence outer scale L_N is $1 < L_N < 5$ m, once again much less than hundreds of meters. In those years, we had no clues about such small temperature outer scales, when compared to velocity outer scales.

In 1994, Vernin and Muñoz-Tuñón were the first authors to give a phenomenological explanation from the analysis of balloon-borne profiles of temperature. They discovered that, within the thickness of a dynamical turbulent layer, due to the rapid Kolmogorov's cascade, the temperature homogenizes, leading to $d\theta/dz = 0$. This example was further detailed by Coulman et al. (1995) and extended to many other cases, where pairs of thin "laminae" appear at each side of a thick dynamical turbulent layer. Almost at the same time, Dalaudier et al. (1994), showed the evidence of "sheets" in the atmospheric temperature field. Five years later, Werne and Fritts (1999), confirmed our phenomenological model using 3D direct numerical simulations.

2.2.1 Phenomenological Model: Laminae

Before going deeper in the model details, let us see examples of vertical profiles of C_N^2. In Fig. 2.9 one can see the fine structure of the optical turbulence. Balloon flight begins at 2.2km, the altitude of the observatory. Many "lines" are visible and each corresponds to a different turbulent layer. Each optical turbulent layer seems to be very thin, much thinner than the hundreds of meters which characterize the outer scale of velocity motion (see Barletti et al., 1974, and, more recently Vernin and Muñoz-Tuñón, 1992).

Now, let us make an altitude zoom between 4 and 5 km of C_N^2 profile of Fig. 2.9, which is plotted next in Fig. 2.10. One can note the fine structure of the optical turbulence layers. Their thickness at half maximum is about ten meters.

In Fig. 2.11, another pair of turbulent laminae is visible for which it was possible to compute the Richardson number as defined in Equation 2.18. In (a) C_N^2 and θ and in (b) R_i are plotted between 6.5 and 8.4 km. In the region where $\Delta\theta/\Delta z \sim 0$ is between the two laminae, the Richardson number is well below 1/4 and close to 0, as predicted in Section 2.1.2.1.

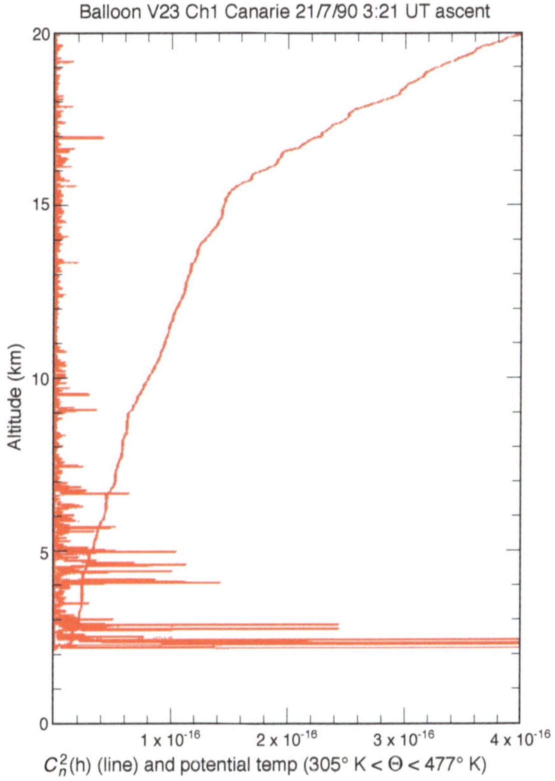

Balloon V23 Ch1 Canarie 21/7/90 3:21 UT ascent
Altitude (km)
0
5
10
15
20
1 x 10^-16
2 x 10^-16
3 x 10^-16
4 x 10^-16
C_n^2(h) (line) and potential temp (305° K < Θ < 477° K)

Figure 2.9 Example of C_N^2 fine vertical turbulence structure. The curve is the potential temperature $\theta(h)$. Flight 23 launched at Observatorio Roque de los Muchachos, La Palma, Canary Island.

In Fig. 2.12 (see also Vernin, 2002, and Avila and Vernin, 1999) we give our phenomenological interpretation of the presence of thin optically turbulent laminae which appear in pairs most of the time. In (a) is sketched the first step, where a dynamical turbulent layer is "growing", but where the wind speed gradient is not

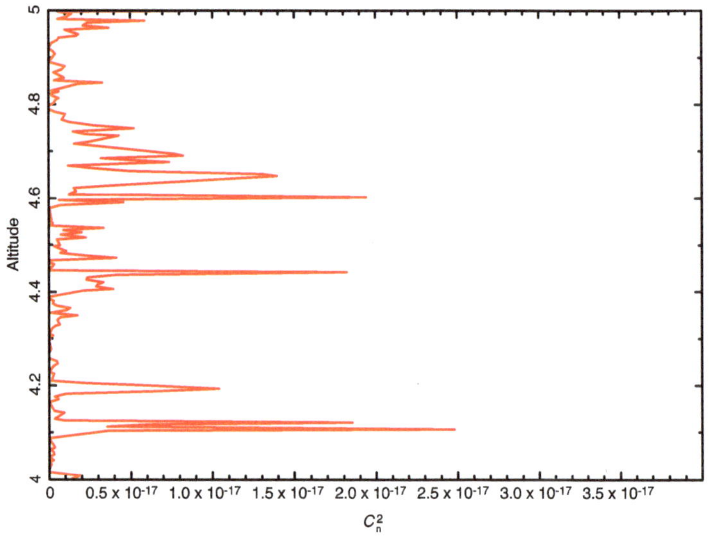

Figure 2.10 Same as Fig. 2.9, but zoomed between 4 and 5 km. Flight 23 launched at Observatorio Roque de los Muchachos, La Palma, Canary Island.

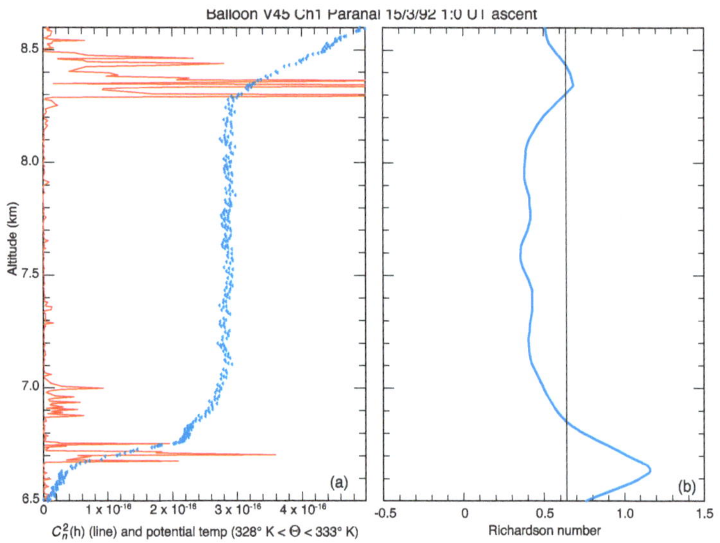

Figure 2.11 Flight 45 launched at Paranal. In (a) C_N^2 and θ appear with altitude, and in (b) $R_i(h)$. Here is the evidence of a fully developed dynamical 1800 m-thick turbulent layer, and two much thinner optical turbulent layers. R_i drops well below 1/4 within the fully developed turbulence zone.

sufficient to overcome the static stability and the Richardson number is greater than 1/4. In (b) is represented a hypothetical state where the wind speed gradient is sufficient enough to break the flow into turbulence, and the Richardson number drops below 1/4. A full Kolmogorov cascade is well established and potential temperature takes any value between a minimum and a maximum that correspond to the bottom and top of the layer. One expects that the optical turbulence reaches its maximum in between the whole turbulent layer, i.e., hundredth of meters, a situation which is never encountered in our real atmosphere. The "normal" situation is presented in (c) where the potential temperature "rapidly" homogenizes, due to thermal conductivity at small scale (inner scale), leading to zero potential temperature gradient. Richardson number reaches zero as well as C_N^2.

The question is "why is case (b) of Fig. 2.12 never encountered?". As we already suggested, (Vernin, 2002), we think there is a step which is forgotten in the above description. Before case (a), at the beginning, a shear flow is developing over a

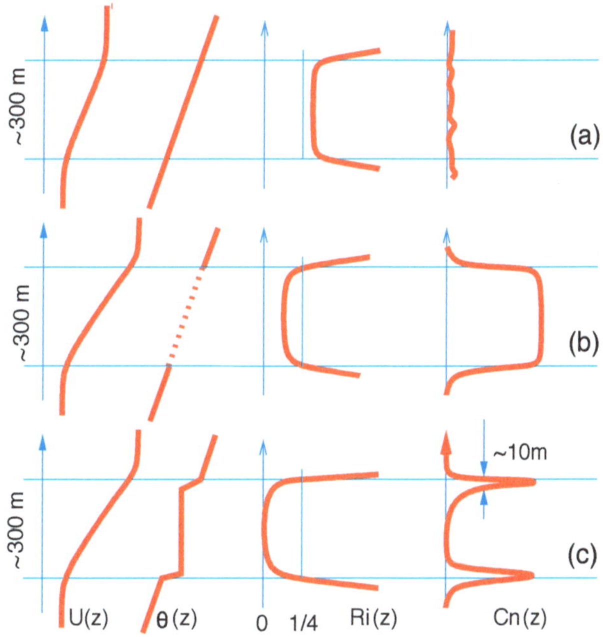

Figure 2.12 Three situations of occurrence of twin optical turbulent laminae: (a) before turbulence onset the Richardson number is larger than its critical value, (b) the wind speed gradient is sufficient to induce dynamical turbulence throughout the whole layer, well mixing the parcels of air with different potential temperature leading to a "never observed" case of a thick optical turbulent layer, (c) rapidly potential temperature homogenizes to reach $\Delta\theta/\Delta z \sim 0$, which induces $R_i = 0$ and $C_N^2 = 0$.

thickness much less than, say, 300 m. A thin dynamical and optical turbulent layer appears. Quickly, due to the dissipation of the heat, the core of this layer tends to homogenize, leading to the formation of two very close laminae. Then, within this slab $d\bar{\theta}/dz \sim 0$ and thus no kinetic energy is required to move up and down a parcel of air. Since we assume that the process is stationary ($d(\varepsilon, E_{K}, E_{pot}...) = 0$) kinetic energy tends to migrate toward both upper and lower parts of the turbulent slab, enlarging the thickness of this slab. As a consequence, the wind shear tends to vanish since in $S = \Delta U/\Delta z$, the ΔU remains constant but Δz enlarges. After a while, $S = \Delta U/\Delta z$ will become too weak to maintain $Ri < 1/4$ and optical turbulence will remain as a "fossil" turbulence and then disappear.

Depending upon the cause that created a wind shear strong enough to trigger turbulence (see Section 2.3), a new shear will appear, fed by meso-scale motions.

2.2.2 Werne and Fritts Numerical Simulation

In 1999, Werne and Fritts simulated a 3D "Kelvin-Helmholtz" instability and showed the development of two laminae at the edge between the turbulent and the laminar flows. In Fig. 2.13, one can see the temporal evolution of the temperature and wind velocity profiles for a simulation made with $Ri = 0.05$. The vertical wind speed profile is set to $u = U_0\tanh(z/h)$, where U_0 is the velocity difference between top and bottom of the layer of thickness h. Temperature increases linearly $T = \beta z$ where β is the mean gradient. Normalized time is given in U_0/h units. One can see that temperature variance is concentrated within one thin layer at time $t = 1$, then rapidly, at time $t \geq 2$, two laminae are observed. For what concerns the horizontal velocity field, the vertical thickness of the layer is about h at time $t = 0$, and increases up to $6\,h$. Another conclusion of this simulation is that the vertical separation of the two laminae is increasing with time.

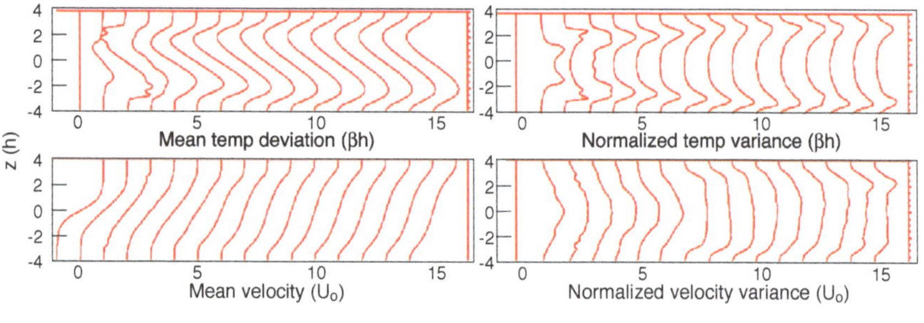

Figure 2.13 Temporal evolution, x-axis, of (top-left) the mean temperature deviation, (top-right) the normalized temperature variance, and of (bottom-left) mean velocity (bottom-right) normalized velocity variance. From Werne et al. (2005).

Most of our observations and related phenomenological model are very coherent with the Direct Numerical Simulation/Large Eddy Scale (DNS/LES) simulations of a stratified turbulent flow.

2.2.3 Optical Turbulence Outer Scale

As a consequence of the concentration of optical turbulence within thin laminae, one cannot expect that the temperature (or refractive index) outer scale L_θ to be of the same order of magnitude as the dynamic outer scale L_0. The L_θ is expected to be much less than the thickness of the laminae, i.e., $\ll 10$ m.

There are many ways to introduce a cut-off in the temperature spectrum. In the literature, one can find in Agabi et al. (1995), references to von Karman or Greenwood-Tarazano cut-off; none of them are based upon physical process, but merely upon mathematical grounds. The value of the spatial coherence outer scale $\pounds_0$, which is related to L_θ, is controversial because the available measurements provide scattered values ranging from a few meters to kilometers (Avila et al. 1997), likely due to the fact that $\pounds_0$ is model dependent. Abahamid et al. (2004) assumed the validity of Eq. 2.91 to deduce L_θ from the knowledge of M and C_N^2. Based upon a set of 168 instrumented balloons, Abahamid et al. (2004) gave the trend of L_θ within the boundary layer (Fig. 2.14) and the free atmosphere (Fig. 2.15). One can see that the optical turbulence outer scale is well bounded according to $0.3 < L_\theta < 4$ m. This conclusion seems very coherent with our phenomenological description of the optical turbulence appearing within thin laminae.

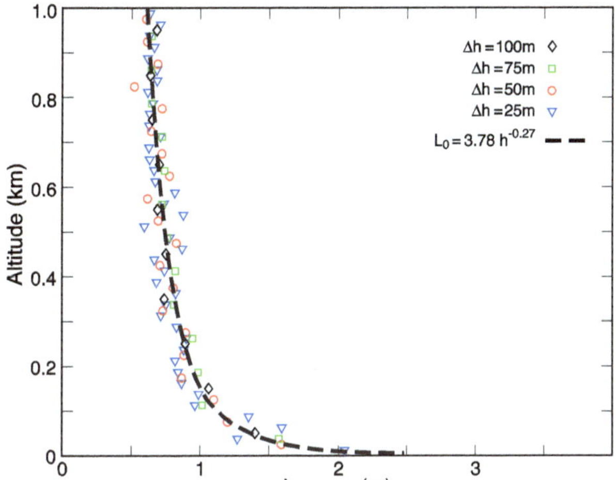

Figure 2.14 L_θ vertical profile calculated for different layer thickness Δh, in the boundary layer. Right: mean profile; left: median profile; dotted-fat line: fitting function.

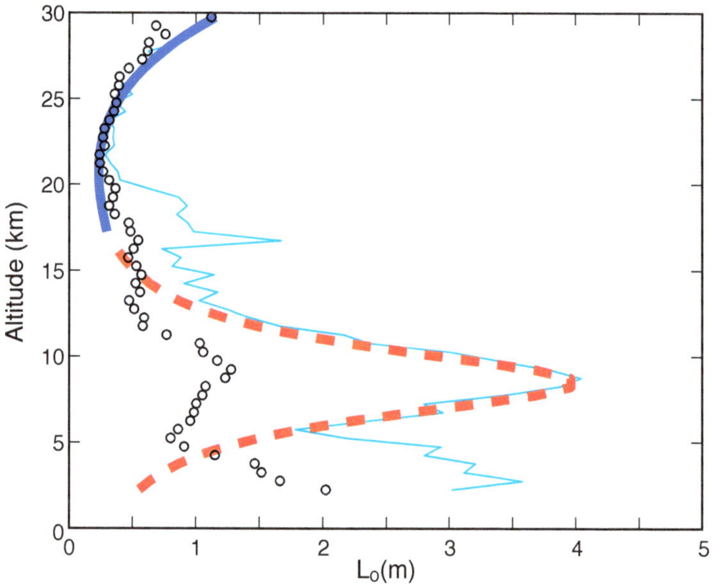

Figure 2.15 Comparison of L_θ vertical profile deduced from our soundings data and the result of the Brown-Beland model (solid fat line) and Coulman-Vernin models (dotted fat line). o: median, solid line: mean.

2.3 Optical Turbulence Generation by Gravity Waves

During night conditions, most of the time our atmosphere is stable ($d\theta/dz > 0$) and kinetic energy is not sufficient to trigger optical turbulence $R_i > 1/4$. In our opinion, the presence of optical turbulence is due to the crossing of inertial gravity waves (GW), which locally increases the wind shear leading the Richardson number to cross the 1/4 threshold. Gravity waves appear in our atmosphere under favorable conditions such as "jet stream" orography or cold/warm air front. A comprehensive description of gravity waves is given by Gill (1982). In 2002 and 2004, two observing campaigns were organized (with the Air Force Office Scientific Research, USA) at "Haute Provence Observatory", France during which many instrumented balloons were launched, as well as observations with remote optical monitors like the Scidar (see Part II of this book). According to Vernin et al. (2007) and Jumper et al. (2007), there is a clear relationship between optical turbulence and gravity waves.

Within assumptions given in the previous articles, the velocity perturbation due to a gravity waves is a vector u, v, w which can be expressed as follows:

$$u = u_0\cos(kx + ly + mz_\omega t) \tag{2.93}$$

where k, l, m are the three spatial wavenumbers of the GW at point x, y, z, and ω is the temporal intrinsic pulsation. Wavenumber and intrinsic pulsation are related through the dispersion equation:

$$\omega^2 = (f^2m^2 + N^2(k^2 + l^2))/(k^2 + l^2 + m^2). \tag{2.94}$$

where N is the Brunt-Väisälä frequency, or buoyancy frequency, defined by:

$$N^2 = \frac{g}{\rho_0} \frac{d\rho_0}{dz} \tag{2.95}$$

and f the inertial frequency at latitude φ:

$$f = 2|\Omega|sin(\varphi) \tag{2.96}$$

and where $|\Omega| = 7.292 \ 10^{-5} \ s$ is the rotation rate of the earth. In Fig. 2.16 one can see a sketch of the (u, v) velocity perturbation. It becomes clear that, if u_0 and v_0 and the eccentricity of the ellipse are sufficient, the Richardson number might cross the 1/4 threshold and give rise to optical turbulence.

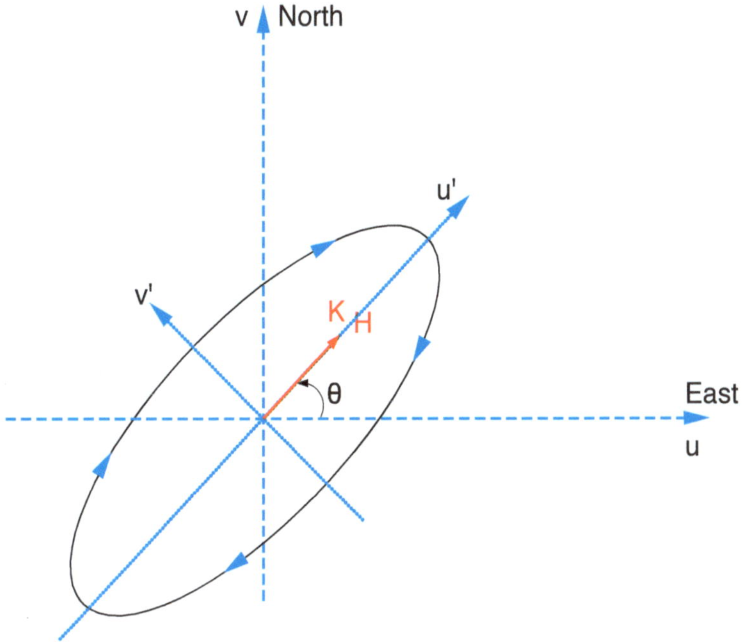

Figure 2.16 Hodograph of (u, v) wind velocity GW perturbation.

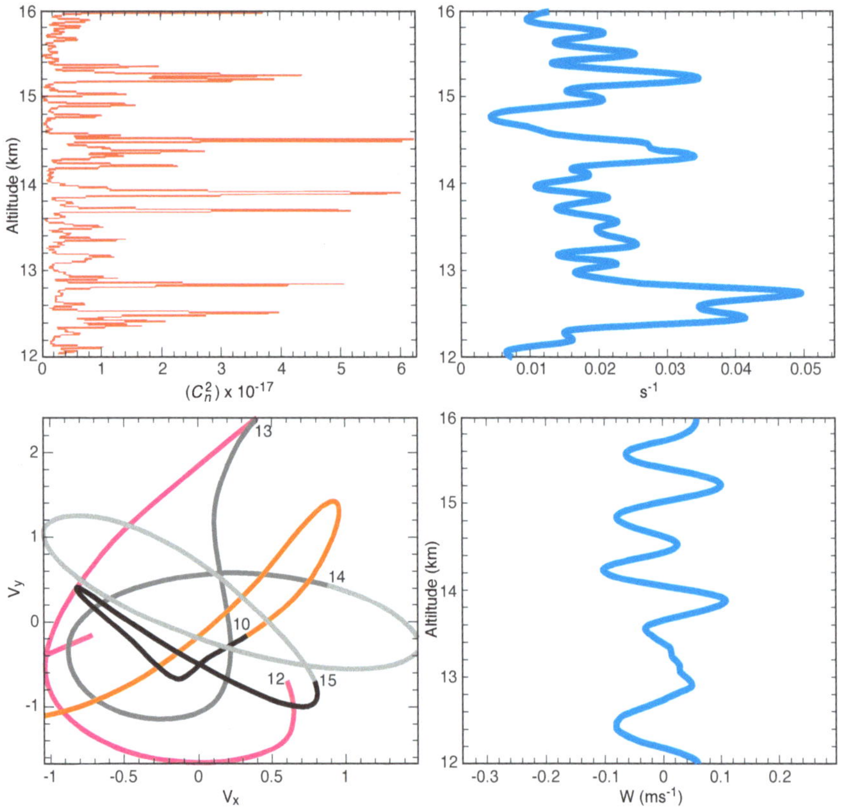

Figure 2.17 Analysis of flight 267, on July 18, 2002. The $C_N^2(h)$, vertical gradient of horizontal wind, wind speed hodograph and balloon ascent speed are plotted from top left to bottom right.

In Fig. 2.17, we detail the GW/OT correlation from the analysis of a particular balloon flight. Considerable GW activity is visible in the hodograph. Again, in the upper part of the plot, optical turbulence layers clearly appear where the vertical gradient is at a maximum.

In summary, in this chapter the concept of optical turbulence, as distinct from dynamic turbulence, was introduced and explained in detail. Basic concepts such as the classical theory of dynamical turbulence, atmospheric stability, and the formation of turbulent eddies (e.g., Kolmogorov, 1941) was presented first, followed by a discussion of the spatio-temporal fluctuations of variables that impact air density, such as temperature (Obukhov, 1949; Yaglom, 1949). The laws governing the spatio-temporal fluctuations of the refractive index were also described. In turbulent layers, the potential temperature gradient is mixed out, with the result that the refractivity profile is characterized by a pair of optical turbulent "laminae" that appear at the margins of the layer. The chapter closes with a short description on the relationship between optical turbulence and gravity waves (Vernin et al., 2007; Jumper et al., 2007).

Authored by
J. Vernin
Laboratoire Fizeau, Université de Nice-Sophia Antipolis,
Observatoire de la Côte d'Azur,
CNRS-UMR6525, 06108 Nice Cedex 2, France

Part II
Instrumentation for Observing
Optical Turbulence

CHAPTER 3

Remote Optical Turbulence Sensing: Present and Future

3.1. General Perspective

The refractive index of atmospheric air is non-homogeneous and distorts propagating light waves. This phenomenon, called *optical turbulence*, causes serious limitations in astronomy, optical communication, adaptive optics, etc., driving the need for turbulence characterization. At the same time, these distortions provide means to measure the optical turbulence by a *remote optical turbulence sensor* (ROTS). Compared to direct *in situ* techniques such as micro-thermal probes, a ROTS has obvious advantages: it can sample the whole turbulent path instantaneously and measures the *optical* effect, directly related to the propagation, hence to the needs of the end users. Unfortunately, the information on turbulence delivered by existing ROTSs is always incomplete due to the limitations of these techniques.

Figure 3.1 gives an overview of a general ROTS and its major components.

The light source must be located outside the turbulent volume. Bright stars (one, two, or several) are most convenient ROTS sources available at no cost and distributed over the whole sky. Extended sources (Sun, Moon, or planets) can be used as well. The uniqueness of these solar-system bodies pre-selects the light path (a ROTS can only work when the source is visible) and does not permit sampling the atmosphere whenever and wherever we want. In this sense, artificial laser guide stars (LGS) are ideal, being created at will where needed. The cost of an LGS is,

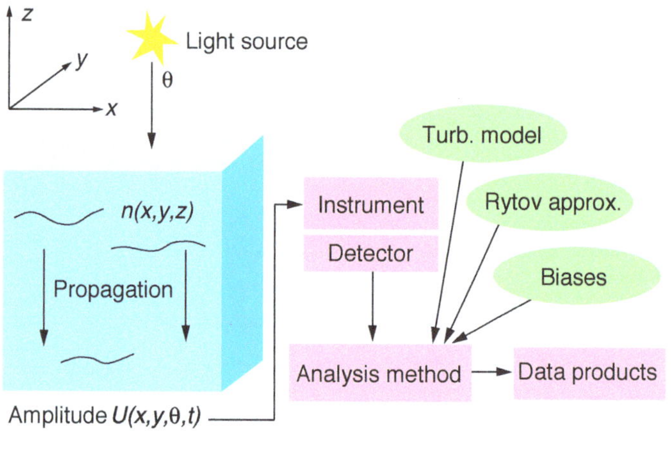

Figure 3.1 Remote optical turbulence sensing

however, prohibitively high for turbulence monitoring (primarily because of the high laser power needed to get enough back-scattered photons), so these sources are suitable for a ROTS only when they are created for another purpose, such as adaptive optics (AO). On the other hand, low-power laser beams are convenient for turbulence sounding on a horizontal path. In principle, an artificial light source can be placed on a mast, airplane, or satellite to serve for a ROTS. For example, a 30-m solar reflector on a geo-stationary satellite can permanently illuminate one hemisphere, creating a near-stationary bright "star" for ROTS.

Optical turbulence can be mapped as an instantaneous 3D distribution of refractive index fluctuations $\Delta n (x, y, z, t)$ by means of turbulence tomography, using several stars (e.g., Tokovinin and Viard, 2001). This is needed in AO for wide-field image correction, but would be an over-kill for turbulence characterization (too much information). Instead, a random-process model is assumed. Local turbulence strength is measured by the C_N^2 parameter (*refractive index structure constant*), implying that the refractive-index spatial spectrum obeys the Kolmogorov law between the *turbulence inner scale* l_0 and the *turbulence outer scale* L_0 (Tatarskii, 1961; Roddier, 1981). A frozen-flow assumption permits describing the temporal evolution of turbulence by the wind speed vector $\mathbf{V}(\mathbf{x})$, where $\mathbf{x} = (x, y, z)$ is a 3D coordinate vector in the turbulent volume.

The random-process model has its obvious limitations. We are dealing with a unique realization of a process that is not stationary in space and time (the turbulence is "patchy", *intermittent*). So, a "local" and/or "instantaneous" C_N^2 $(\mathbf{x},t)$ value implies some spatial and/or temporal averaging and can be defined or measured only approximately. Atmospheric quantities typically vary by several orders of magnitude, so geophysicists are happy to measure them to within a factor of two, in which case the random-process model is adequate. A higher accuracy is sought in optical propagation applications, so, eventually, the limit will be set by the model, rather than by the measurement errors. Until then, these are the primary *data products* of a ROTS:

1. *turbulence profile* (TP) $C_N^2 (z,t)$ (it is usually assumed that turbulence is stratified horizontally, with no dependence on x,y);
2. *wind profile* $\mathbf{V}(z)$ (same assumption), $\mathbf{V} = (V_x, V_y)$;
3. *outer scale profile* $L_0(z)$,

and the derivatives of these quantities include seeing ε_0, isoplanatic angle θ_0, AO time constant τ_0, etc. (see Preface). The inner scale l_0 is almost always irrelevant to optical propagation in astronomy.

Propagation of light through a turbulent volume is such a complex phenomenon that there is no exact theory. Instead, a small-perturbation (Rytov) approximation (Tatarskii, 1961; Roddier, 1981) is universally used to relate turbulence parameters with the lightwave statistics. Depending on the conditions (turbulence intensity, spatial scale, wave- length), this approximation can be either excellent or poor. Fortunately, modern computing power permits solving the wave propagation numerically and hence evaluating or extending the validity of the canonic Rytov theory.

Instrument receives the light waves distorted by the atmosphere and performs some optical transformations. For example, the image of a star or the Sun can be formed at the focus of a small telescope. Alternatively, we can study the distribution of the light amplitude or phase at the pupil plane. The number of potential transformations is unlimited. In the end, the information on the light wave distortion is encoded as a light intensity, and the light is detected by a CCD or some other device.

Analysis method is closely related to the instrument design. The intensity of light in individual detector pixels must be processed, related first to the statistics of the wave distortions and then to the turbulence parameters. Intrinsic limitations or biases, such as photon noise or finite exposure time, must be accounted for. Moreover, the departure of the real instrument from its ideal model can critically affect the results. Suitable methods to discover or monitor instrument parameters and data quality are an essential part of the analysis.

Data products of any given ROTS are either primary quantities (turbulence and wind profiles) or their derivatives such as seeing. Each measurement refers to a particular moment in time and a particular viewing direction. Comparisons between different instruments or between a ROTS and a telescope usually involve different directions, hence are subject to non-stationarity errors.

3.2 History

The foundations of the propagation theory and turbulence models were set in the 1950s (Chernov, 1960; Tatarskii, 1961). However, quantitative ROTS appeared much later. Good reviews of the basic physics and early techniques are provided by Roddier (1981) and Coulman (1985). Progress is mainly determined by the available technology, and current rapid development of the ROTSs is a reflection of our ability to capture and analyze the light better.

Astronomers took time to assimilate and use the advances of the turbulence theory. The first seeing monitors with automatic photo-electric registration of the absolute or differential image motion appeared in the 1960s; and now the Differential Image Motion Monitor (DIMM) is a well-established standard instrument (Sarazin and Roddier, 1990; Tokovinin, 2002b). The idea of DIMM can be traced back to Hosfeld (1954), DIMMs with visual detection were developed by Stock and Keller (1961), a photoelectric DIMM was apparently pioneered by Gillingham (1978). Absolute image motion was first detected photographically with Polar-star trail techniques (Harlan and Walker, 1965), while Babcock (1963) had already measured it photo-electrically.

In the 1970s, the Astrophysical Department of the University of Nice (led by F. Roddier) was at the forefront. Basic ideas of measuring optical turbulence by means of wavefront slopes or scintillation were formulated and tested. The latter method, SCIDAR (SCIntillation Detection And Ranging) (Rocca et al., 1974; Vernin and Azouit, 1983), can deliver turbulence and wind-speed profiles by using a double-star source. It remains one of the most powerful ROTSs. The implementation of SCIDAR, now almost trivial, became possible only with an ingenious real-time image correlator developed by M. Azouit. It was still a pre-computer era!

Accessibly cheap computers and image detectors finally appeared at the end of the 1980s. As both of these critical elements improved their performance, so did the ROTSs: DIMMs became faster, SCIDARs cheaper. A Generalized Seeing Monitor (GSM) was the first instrument to measure turbulence outer scale on a routine basis using an old idea of angle-of- arrival correlation (Ziad et al., 2000). Interestingly, GSM employed a now-obsolete image modulation technique (the CCDs were still not fast enough). Yet another step in the ROTS development looks trivial, but turned out to be crucial: the DIMMs became robotic and started to provide a continuous data flow, first at European Southern Observatory (ESO), then at some other observatories. Even today, many ROTS instruments have not passed this milestone and continue to be manually operated, one-time experiments rather than monitors.

The development of ROTS instruments continues in the current decade (2000s) at a fast pace, driven by the increased demand, cheaper components, more computer power, and new ideas.

3.3 Current Techniques

There are an infinite number of ways to derive turbulence characteristics from the statistics of distorted light waves. Which is the best? Where are the limits? Let us try to inter-compare current methods and their potential improvements in trying to answer these questions. The methods are divided in broad classes according to the measurement principle.

3.3.1 Long Exposures

This is a class of methods where the light intensity is averaged over a long time (much longer than the aperture transit time D/V, e.g., for 30 s or more). The effect of turbulence on long exposures is to reduce the effective transverse coherence length of the field to the size of the Fried parameter r_0. The Point Spread Function (PSF) becomes wider, about λ/r_0 rather than diffraction limit λ/D (D–diameter of the aperture, λ–wavelength of light).

The size of point-source (stellar) long-exposure images is a practical and widely used measure of "seeing". The problem of this method is that the PSF is also affected by guiding errors and telescope aberrations, including defocus. We thus deduce from the PSF an upper limit on seeing. At modern telescopes with good optics and guiding, this measure is close to the atmospheric seeing. For example, the best long-exposure images at VLT have FWHM size of 0.2″, so we may be confident that a FWHM of, say, 0.5″ is essentially produced by the turbulence. Part of this turbulence can be inside the dome or on the telescope mirror.

The problem of unknown optical aberrations has been avoided elegantly by measuring the fringe contrast in the pupil plane with a *coherence interferometer*. Atmospheric aberrations are variable and degrade the contrast of long-exposure fringes, whereas optical aberrations are static and only change the average fringe phase. These interferometers superpose two wave-fronts with a relative rotation (Roddier and Roddier, 1973) or mirror flip (Dainty and Scaddan, 1975). The baseline r changes across the pupil, and so does the contrast of long-exposure fringes $\gamma(r)$,

$$\gamma(r) = \exp[-3.44(r/r_0)^{5/3}]. \tag{3.1}$$

This method is theoretically perfect, but still affected by the guiding errors. Yet another problem is that Eq. 3.1 is accurate only for an infinite outer scale L_0. A realistic outer scale $L_0 \sim 20$ m affects the coherence even at baselines of $r \sim 0.1$ m. The method still works, because we measure the true coherence (hence the true PSF), but the r_0 estimate derived from Eq. 3.1 will always be biased to larger values unless L_0 is known. The influence of the outer scale can be used to advantage: by measuring the long-exposure PSF (or coherence) at optical and infrared wavelengths simultaneously, we can derive the pair of parameters (r_0, L_0) (Tokovinin et al., 2007).

3.3.2 Absolute Image Motion

Absolute tilt of a wave-front over a circular telescope aperture (also called image motion or angle-of-arrival) is related to the seeing as

$$\sigma_\alpha^2 = K(\lambda/D)^2 (D/r_0)^{5/3}, \tag{3.2}$$

where the coefficient $K = 0.340$ if the tilt is defined as an average wave-front gradient, and $K = 0.360$ if the Zernike tilt (best-fitting plane) is meant. The coefficient K is different for a centrally obstructed or a non-circular aperture (Sasiela, 1994). The motion of an image in the focal plane of a small telescope thus leads to the estimate of the σ_α^2 and seeing, r_0.

Both Eq. 3.2 and Eq. 3.1 are valid only for the Kolmogorov turbulence model, $L_0 = \infty$. The measured tilts can be corrupted by the telescope shake. So, the absolute-tilt methods and long-exposure techniques are similar. The advantage of tilt vs. PSF is that the aperture diameter D can be smaller than r_0. The problem of telescope shake can be avoided by firmly fixing the telescope during data acquisition. Of course, the star moves in the field, but this linear trend can be fitted and subtracted, leaving pure atmospheric signal, or we can measure only in the direction perpendicular to the diurnal motion. The Polar Star is the source of choice for the northern hemisphere, as it moves so slowly. Polar star motion was recorded both photographically (Harlan and Walker, 1965) and photoelectrically (Babcock, 1963; Scheglov, 1980). In the southern hemisphere, the telescope has to track the star, so the measurements are usually done in the declination direction, as in the Carnegie seeing monitor (Persson et al., 1990).

The sensitivity of absolute image motion to L_0 is exploited in the Generalized Seeing Monitor (GSM) to measure this parameter (Ziad et al., 2000). The outer scale is deduced from the covariance of the image motion in 4 telescopes on individual mounts, with the idea that tracking errors (in the declination direction) are mutually independent and thus do not bias the covariances. As an additional precaution, parabolic trend is subtracted to model slow drifts. The r_0 parameter is measured independently by the differential motion (see below). The covariances are normalized by $r_0^{-5/3}$ to remove their dependence on seeing and fitted to the model depending only on L_0.

To the first approximation, the image motion is proportional to the phase difference at the aperture borders. Phase difference over a large baseline can be measured also by an interferometer. The influence of finite outer scale L_0 is even greater at large baselines, so interferometric data can be used to measure it, as long as an independent estimate of r_0 is available. A comparison of the GSM with a long-baseline interferometer demonstrated a good agreement between these two approaches (Ziad et al., 2004). In fact, the GSM was compared to the speed of the fringe motion in the interferometer which, of course, is proportional to the product of the wave-front tilt and the wind speed. It comes as no surprise that tilts measured by the interferometer and the GSM agree well.

3.3.3 Differential Image Motion

In a DIMM, the seeing is inferred from the variance of the differential wave-front tilts in two (or more) small apertures. Typically, circular apertures of diameter D with centers separated by a baseline B are used. The variance of the differential tilt is described by Eq. 3.2, but the coefficient K now depends on the B/D ratio, on the measurement direction (longitudinal is parallel to the baseline, transverse is perpendicular), and on the details of centroid calculation (Tokovinin, 2002b). Moreover, the DIMM response is affected by the optical propagation (neglected in the standard DIMM theory) and by optical aberrations (Tokovinin and Kornilov, 2007). So, a blind application of Eq. 3.2 would not give an accurate r_0 estimate unless additional care is taken to remove instrumental biases.

A DIMM can be considered as a simplest Shack-Hartmann *wave-front sensor* (WFS) with two sub-apertures. To the first approximation, differential tilts are related to the defocus and astigmatism aberrations over a full telescope aperture, i.e., the wave-front curvature. A WFS with more than two sub-apertures can measure these and higher order aberrations. Any *adaptive-optics* (AO) system with such a sensor can be used to deduce the seeing. Indeed, most AO systems have a seeing estimator incorporated in their software (e.g., Fusco et al., 2004; Schöck et al., 2003). The advantage of this solution is that the seeing is measured on the optical path that is relevant for the AO itself, avoiding any assumptions about turbulence spatial distribution or stationarity.

To be fair, estimating the seeing and other atmospheric parameters from the AO loop data is not a simple matter. The response of a WFS is often non-linear and poorly calibrated. This does not matter much for the AO itself, as the WFS measures only residual aberrations of the compensated wave-front. In a closed loop, the atmospheric distortions are derived from the shape of the *deformable mirror* (DM), which, in turn, is deduced from the control signals (voltages) and from the more-or-less accurately known properties of each particular DM. The temporal response of the closed-loop AO must be taken into consideration, noise variance must be subtracted, etc. Accurate seeing measurement is never a high priority for any given AO project, so atmospheric parameters currently measured by AO systems should be taken with a grain of salt until detailed analysis of biases and inter-comparisons with other methods are done.

Using a single star, we can only measure integrated turbulence properties along the line of sight. With a double source, it is possible to separate different atmospheric layers using the principle of *crossed beams* (Fig. 3.2). This idea is so universal that it is found in various ROTSs and is also used on horizontal paths with crossed laser beams. In the region of the beam overlap, turbulence gives correlated signals, while on the remaining part of the path the signals are un-correlated. Simple geometry tells us that the sensed zone is at a distance $z = B/\theta$ from the receiver, where B is the distance between the beams at the receiver and θ is the small angle between the beams. We can measure the turbulence profile on the whole path by changing B, θ, or both. The resolution along the line-of-sight will be $\Delta z = d/\theta$, where d is the beam diameter.

One popular incarnation of the crossed-beam principle is SLODAR (SLOpe Detection And RAnging) (Wilson, 2002). It uses the double star with a fixed angular distance θ between the components as a light source. The wave-fronts are measured separately on each star using a standard WFS, usually (but not necessarily) of a Shack-Hartmann type. The number of resolution elements along the path equals the number of sub-apertures d across telescope diameter D, $N = D/d$. The sensing range extends from the telescope aperture to $z = D/\theta$ and depends on the separation θ of the chosen double star. The resolution is $\Delta z = z/N = d/\theta$. In a SLODAR, the average tilt over all apertures is subtracted and only the differential tilts are correlated. This makes the method immune to the telescope shake, but complicates data interpretation. To get a turbulence profile, the measured correlation between differential tilts of two stars is de-convolved with either theoretical or experimental tilt auto-correlation function as a kernel (Butterley et al., 2006).

Although SLODAR uses a standard WFS, it is a separate instrument, not part of an AO system. Care is taken to calibrate the WFS properly and to ensure un-biased slope measurements, marking a difference with current AO systems. Slope is easy to measure for sub-apertures of ~20 cm or larger because suitably bright double stars can be found, while the exposure time can be shorter than the sub-aperture crossing time d/V. However, it was possible to build a portable SLODAR with 5-cm sub-apertures by using a state-of-the art CCD detector with internal gain. This instrument has $N = 8$ sub-apertures on a 40-cm telescope.

The potential of the SLODAR method is not yet fully explored. It can measure a low-resolution turbulence profile in the whole atmosphere with a narrow double star or, alternatively, can use a very wide stellar pair for detailed, high-resolution profiling of the ground-layer turbulence. It is possible to extend the method to three or more stars. Multiple guide stars are necessary anyway for turbulence tomography in AO, so the signals from multiple WFSs will be available for SLODAR turbulence profiling (see also Tokovinin and Viard, 2001, for an alternative profiling method).

Another variation on the crossed-beam theme is a group of ROTSs where the baseline B is fixed, and a multitude of sources with different angles θ is observed. The edge of the solar or lunar disks is such a convenient multi-θ source. Differential motion of the solar limb is a standard method of daytime seeing measurement for solar astronomy (Beckers, 2001), although a DIMM can work in daytime with bright stars, too. The Moon is bright enough for night-time measurements. Recently, an experiment to measure differential tilts of the lunar limb, called MOSP (Moon Outer Scale Profiler), has been performed (Maire et al., 2007). So far, MOSP used only one full aperture. Differential tilt along the limb was measured and modeled. These tilts are sensitive to the turbulence outer scale L_0, so MOSP can, in principle, estimate the vertical profile $L_0(z)$. In the current implementation, an independent measurement of $C_N^2(z)$ is needed to model the data, but MOSP with two or more sub-apertures will be a self-sufficient ROTS that will measure both $C_N^2(z)$ and $L_0(z)$.

3.3.4 Scintillation

Twinkling of stars, *scintillation*, is caused by the optical turbulence and therefore can serve to characterize it. It is easier to measure flux than to measure wave-front tilts or phases. The optical propagation itself converts phase fluctuations to intensity variations, replacing "Instrument" in the general ROTS scheme (Fig. 3.1). All we need is just a detector!

The scintillation strength is characterized by the scintillation index σ_I^2,

$$\sigma_I^2 = \langle \Delta I^2 \rangle / \langle I \rangle^2, \tag{3.3}$$

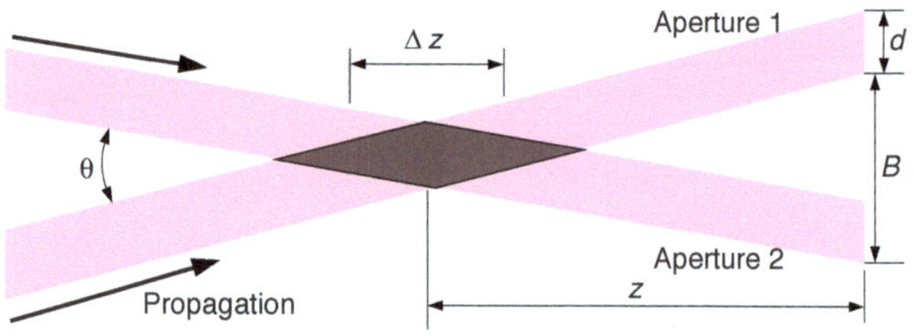

Figure 3.2 Principle of a crossed-beam ROTS. Two beams propagate from left to right toward receiving apertures. The dark area shows the beam overlap zone.

where I is the instantaneous light intensity. Theory operates with the intensity distribution $I(x,y)$, in practice I is always averaged by some finite receiving aperture d or detector pixels.

The scintillation (speckle) pattern $I(x,y)$ (also called *flying shadows*) has a characteristic scale of the Fresnel radius $r_F = \sqrt{\lambda z}$, z being the propagation distance. Details larger than r_F can be described by geometric optics, as result of random focusing. Hence, for large apertures $d \gg r_F$, $\sigma_I^2 \propto z^2$. On the other hand, $\sigma_I^2 \propto z^{5/6}$ for the intensity fluctuations at a given point, or for small apertures $d \ll r_F$ (Roddier, 1981). Strong dependence of scintillation on z prevents us from interpreting scintillation index in terms of seeing unless the distance z is known. This means that a scintillation-based ROTS should measure somehow the turbulence profile (TP) $C_N^2(z)$. Yet another complication is the approximation of weak turbulence $\sigma_I^2 \ll 1$ used always in the data analysis but not always valid in real life.

The first such instrument, Scintillation Detection and Ranging (SCIDAR), is based on the crossed-beams idea (Rocca et al., 1974; Vernin and Azouit, 1983). A binary star with angular separation θ produces the scintillation pattern where each detail is duplicated with a spatial shift θz. Of course, contributions from different layers are all mixed, but they can be recovered with correlation analysis, by computing the *covariance function* (CF),

$$C_I(\zeta,\eta) = \langle \Delta I(x + \zeta, y + \eta)\, \Delta I(x,y) \rangle / \langle I \rangle^2. \tag{3.4}$$

If the scintillation patterns from two stars were detected separately and cross-correlated, each turbulent layer would produce one peak shifted by θz from the coordinate origin. In a SCIDAR, the light of the stars is mixed together, so each layer produces a peak at the coordinate origin (self-correlation of each star) and a pair of symmetric peaks at $\pm \theta z$ (cross-correlation). The observed function $C_I(\zeta,\eta)$ is a superposition of all such triplets. The profile can be extracted by deconvolving it with the known response function. It is relatively easy to identify and measure peaks produced by strong turbulent layers, but full deconvolution also recovers a smooth component of the TP which is not so obvious but can dominate the overall seeing.

As the width of the CF peak is of the order of r_F, the vertical resolution of SCIDAR is $\Delta z/z = \sqrt{\lambda z}/\theta z$. With a 10″ binary and $z = 10$ km, SCIDAR has a vertical resolution $\Delta z \approx 1.5$ km. The size of the receiving aperture must be no less than θz, or 1 m if we want to reach 20 km with a 10″ binary.

Turbulence near the telescope produces no scintillation. However, in a Generalized SCIDAR (G-SCIDAR) the detector is conjugated to the virtual plane below telescope aperture (Avila et al., 1997; Fuchs et al., 1998). This adds effectively some propagation path and permits to measure the complete TP. This virtual propagation is mathematically correct only when we neglect diffraction on the aperture border and optical aberrations, which is a reasonable approximation for telescopes above 1 m in size and conjugation planes some 2-5 km below the ground. Virtual propagation increases the scintillation index and associated departures from the weak-turbulence regime. All modern SCIDARS work in the generalized mode.

Apart from the TP, a SCIDAR can measure the wind speed in the atmosphere by correlating intensity patterns with some time lag τ. In such *spatio-temporal* correlation the peaks are displaced by $V\tau$. Again, it is easy to measure the speed of prominent peaks, but difficult to treat a general case of smoothly distributed turbulence. Automatic retrieval of both turbulence and wind profiles from G-SCIDAR data has been proposed and tested by Prieur et al. (2001).

If we select a very wide binary, the vertical resolution of a SCIDAR increases, at the expense of the altitude range. The idea of LOw-LAyer SCIDAR, LOLAS, is exactly this (Avila, 2007). Diffraction at the aperture edge becomes even more important, but its effect can be accounted for by modeling. LOLAS has been proposed for detailed profiling of the ground layer and used in the Mauna Kea Gemini campaign in 2007.

One practical disadvantage of both SCIDAR and LOLAS is a need of binary-star sources with suitable separations. As the choice of bright double stars on the sky is limited, these instruments must observe stars as faint as magnitude 5-7 for a quasi-continuous time coverage. Detectors of the highest possible sensitivity are required. The problem is more acute for LOLAS because its resolution element $\sqrt{\lambda z}$ is smaller. In fact, LOLAS uses an electron-multiplication CCD capable of detecting individual photons, similar to the portable SLODAR.

It is frustrating to use faint double stars in a ROTS while so many bright single stars are available on the sky. Can a single star be somehow doubled? One suggestion was to observe the star at two different wavelengths, so that its image is doubled by the atmospheric dispersion, but this approach never worked in practice. On the other hand, temporal cross-correlation is a convenient way to get two identical but shifted patterns emulating a binary star. If the wind speed in the atmosphere is everywhere the same, $V(z) = $ const, the peaks produced by different layers still overlap in the CCF, but otherwise they are separated. This is the principle of the Single-Star SCIDAR (SSS), introduced by Caccia et al. (1987) and implemented recently (Habib et al., 2006). Recovering both wind and turbulence profiles from a set of correlation functions with different time lags τ is a very complex mathematical problem, with possibly non-unique solution. It has been demonstrated that SSS can detect and measure several strong layers in the atmosphere. It may be more difficult to recover a smooth $C_n^2(z)$ profile. The SSS has been proposed as a turbulence monitor working with a small aperture continuously. So far, it produced data only for short periods of time. The instrument itself is simple, but the data processing may be complex.

A simpler approach to turbulence monitoring with a small telescope is implemented in the Multi-Aperture Scintillation Sensor, MASS (Kornilov et al., 2003; Tokovinin et al., 2003). Here we measure the scintillation of a single star through 4 concentric-ring apertures simultaneously. These apertures act as a spatial filter. The fluctuations of the intensity ratio between two such apertures, called *differential scintillation index*, do not have a steep dependence on z, as normal scintillations, but rather saturate for $\sqrt{\lambda z} > d$, where d is the average diameter of the apertures. This observable is directly related to the seeing produced by the high layers. Indeed, MASS gives very reliable measurements of the free-atmosphere seeing. On the other hand, its vertical resolution is only of the order $\Delta z/z = 2$; it is deter-mined by the match between the Fresnel radius r_F and aperture diameters d_i. MASS uses the propagation theory to disentangle contributions of different altitudes by the size of the flying shadows. The idea is quite old (Peskoff, 1968), it was implemented in the scintillometer of Ochs et al. (1976). The success of MASS, compared to its predecessor, is explained by simultaneous 4-channel measurement and a careful treatment of instrumental effects and propagation. Small departures from the weak-scintillation theory are modeled and implemented in the data processing (Tokovinin and Kornilov, 2007). The atmospheric time constant τ_0 can be evaluated approximately by temporal analysis of scintillation in the smallest MASS aperture (Tokovinin, 2002a). MASS is usually combined with DIMM in a single instrument (Kornilov et al., 2007).

MASS does not sense turbulence near the ground. In principle, we can conjugate its apertures to a virtual plane below the telescope, as in a G-SCIDAR. The generalized MASS was tested and found to be impractical for several reasons. One fundamental consideration is that the scintillation produced by the near-ground turbulence in this regime is of fine spatial scale and small amplitude, i.e., difficult to measure. Using some other optical effect like tilt is a better choice. A comparison between LOLAS and SLODAR, two instruments for measuring the ground layer with scintillation and tilts, respectively, shows that the second method should be more sensitive.

Scintillation of extended sources such as planets or Moon is not seen by a naked eye; its amplitude is very small, but still not quite zero. A source of angular diameter θ averages flying shadows over a circle of diameter θz, acting in the same way as a receiving aperture. The combined effect of such averaging with propagation results in the *decrease* of scintillation with distance as $z^{-1/3}$, so that the maximum contribution comes now from very low altitudes $z \sim d/\theta$. With a receiver of $d = 1$ cm diameter and the Sun ($\theta = 0.5°$), the maximum response is reached at $z \sim 1$ m.

The SHABAR instrument uses solar scintillation for measuring near-ground turbulence (Beckers, 2001). Six small photo-detectors are arranged in a linear non-redundant configuration with baselines up to 0.4 m. Pair-wise covariances of these signals give $n(n-1)/2 = 15$ numbers ($n = 6$) and permit reconstruction of the TP over the first 100 m above the observatory. SHABAR was one of the major components in selecting the site for the new solar telescope (Socas-Navarro et al., 2005). This site was not the best in terms of the measured seeing, but it has been proven with SHABAR that the seeing at the height of the telescope tower will be superior over other sites.

A need to make similar assessment for night-time sites (i.e., to translate the seeing measured by a monitor to the level of the telescope) attracted attention to the lunar scintillometers. One such instrument has been constructed by Hickson and Lanzetta (2004), a 12-channel version is actually deployed at Cerro Tololo. Typical scintillation amplitude is small, $\Delta I/I \sim 10^{-4}$ ($\sigma_I^2 \sim 10^{-8}$), but the Moon is still bright enough to measure such signal reliably. Lunar phases complicate the data reduction but otherwise are not a major problem. Even a simplified 4-channel LUnar SCIntillometer, Lusci, can reliably measure the ground-layer seeing and differentiate contributions from several zones within the first 200 m above ground (Tokovinin et al., 2007).

The beauty of lunar and solar scintillometers is in their simplicity and robustness. With an extended source, the effective size of the averaging area (the largest of θz or d) is always larger than the Fresnel radius. Hence, the scintillation is determined by pure geometric optics, it is achromatic and immune to saturation. One caveat is that the averaging scale θz exceeds 10 m for $z > 1$ km, hence the unknown outer scale L_0 changes drastically the high-altitude response of a scintillometer. This effect was found empirically in SHABAR and is called "missing scintillation" (Socas-Navarro et al., 2005). Fortunately, it does not influence the response at low altitudes.

Scintillometers for turbulence measurements on a horizontal path are produced commercially by *Scintech*. By using large emitting and receiving apertures, these instruments work in the same geometric-optics regime as SHABAR, hence they are immune to the saturation and the turbulence inner scale. These effects would become significant if a narrow laser beam were used instead. Applications of scintillometers for measuring local turbulence (e.g., inside domes) can be envisioned, although high sensitivity is reached only on long propagation paths.

3.4 Future Directions

3.4.1 New ROTS Instruments

The inventory of existing ROTS methods shows a large variety of solutions and deliverables. Table 3.1 is not intended as a "shopping list" for a potential user, rather as an indication that many approaches are possible, and many more will come in the future. The most complete information on turbulence is provided by SCIDAR, albeit at a cost of large telescope aperture, high data flux, and computer-intensive data processing. Simpler and cheaper instruments like DIMM or MASS are better suited for continuous turbulence monitoring. Each of the existing instruments can (and must) be improved in terms of data quality, but also simplicity, cost, and ease of operation. For example, it was found only recently that a DIMM is very sensitive to optical aberrations (Tokovinin and Kornilov, 2007).

Table 3.1 Comparison of existing ROTS methods

Method	$C_n^2(z)$	$V(z)$	r_0	θ_0	τ_0	L_0	Notes
Long-exposure PSF	-	-	+	-	-	+?	shake-sensitive
Coherence interferometer	-	-	+	-	-	-	shake-sensitive
Absolute tilts	-	-	+	-	-	+	shake-sensitive
DIMM or WFS	-	-	+	+?	-	-	standard method
GSM	-	-	+	+	+?	+	
SLODAR	+	+	+	+?	+?	-	double star
MOSP	+?	-	+?	+?	-	+	$L_0(z)$
SCIDAR	+	+	+	+	+	-	double star
LOLAS	+	-	-	-	-	-	GL only, double star
MASS	+	-	+?	+	+	-	low resol., FA
Single-Star SCIDAR	+	+	+	+	+	-	More tests needed
SHABAR	+	-	-	-	-	-	Sun or Moon, GL only
FADE	-	-	+	-	+	-	New instrument

It is unlikely that the explosive development of ROTS stops now. The progress of technology will continue to influence the implementation of the existing methods. Equally unlikely is the prospect that all basic methods are already known and will only be improved in the future. On the contrary, the number of ways in which information on turbulence can be extracted from the optical waves is unlimited. The field of ROTS (and optical propagation in general) is still young, so new methods and ideas will appear inevitably.

One example of a new approach is the FAst DEfocus monitor, FADE (Kellerer and Tokovinin, 2007; Tokovinin et al., 2008). It is a descendant of DIMM where two apertures are replaced by a continuous ring. The radius of the ring-like image formed in the focal plane changes due to the atmospheric defocus aberration, so that seeing can be deduced from these fluctuations. Unlike DIMM, FADE is circularly symmetric, therefore temporal characteristics of the focus fluctuations do not depend on the wind direction. It has been shown that the speed of the focus variation is directly related to the atmospheric time constant τ_0, and, in fact, FADE has been developed explicitly for measuring this parameter.

The birth of FADE illustrates several aspects of future ROTSs. First, the instrument answers a practical need to monitor τ_0 with a small telescope (specifically related to the Antarctic site-testing program). Second, it was enabled by the new technology, fast image acquisition with a machine-vision CCD camera. Finally, the interpretation of the signal in terms of τ_0 required some analytical development and numerical simulations. Conceivably, these three ingredients—*practical need, new technology, and advanced theory*—will be present in all future ROTSs as well.

One potentially fruitful approach will be to capture and analyze defocused stellar images in a small (20-30 cm) telescope. Technically, this is simpler than MASS and DIMM, but can potentially provide similar information on turbulence. An ingredient still missing here is an adequate theory.

3.4.2 Sensitivity Limit

Is there a fundamental limit to the sensitivity of a ROTS? Only a very general argument can be given here. Let the spatial size of the measured optical perturbations (aperture, detector pixel, etc.) be d. The time resolution has to be of the order $\tau = d/V$, where V is the maximum wind speed. So, the number of photons in each instantaneous exposure $N \propto d^2/\tau \propto d^3/V$. The relative variance of the flux caused by the photon noise is proportional to $1/N$. Turbulence-induced phase variance is proportional to $d^{5/3}$ and produces some fluctuations of the detected flux in a ROTS—its signal. So, generally speaking, the signal-to-noise ratio should be proportional to $d^{11/3}/V$. The sensitivity is a very strong function of the aperture (spatial resolution) of a ROTS, hence small apertures are restricted to bright light sources.

State-of-the-art detectors with single-photon detection permit making a smaller ROTS, up to a certain limit. For example, a SLODAR with 40-cm aperture cannot be made even smaller, as its limit is already reached. A typical DIMM uses baseline of 20 cm or more because its sensitivity with a smaller telescope would not be adequate. A MASS with 1-cm inner aperture could measure turbulence at 200 m altitude, but only with stars of magnitude zero. Even with current 2-cm inner aperture, the differential scintillation signal in a MASS is often smaller than the photon noise which is, of course, subtracted from the measured variance.

We conclude that many existing ROTSs are already close to their respective sensitivity limits. However, there is no universal limit. Even weak turbulence can be measured with a faint source by using an adequately large aperture and/or by designing an instrument that translates small perturbations to large flux variations. For example a 15th magnitude star used for guiding and astigmatism correction in a 2×2 Shack-Hartmann sensor on a 8-m telescope is probably bright enough to measure the seeing.

3.4.3 Practical Recommendations

Practical difficulties encountered in the implementation of ROTSs are not obvious from the general principles outlined above. Here are some comments to illustrate the point.

Existing seeing monitors use different components and software and often require substantial efforts in maintenance and data quality control. The "cost of ownership" may be quite high. An observatory wishing to install or improve a seeing monitor discovers that there are no trusted commercial suppliers of such instruments. Eventually they copy or buy an instrument from some other observatory. The *need for standard, commercially available ROTSs* is evident. Hopefully, the reduced cost of detectors and computers will help in fulfilling this need. Site-testing equipment procured commercially in the recent years by large-telescope programs is still too expensive. A future site monitor must be a "set-and-forget" item at accessible price, like a standard meteo-equipment.

It turns out that the key missing component is not a ROTS itself but its feeding telescope. Amateur robotic telescopes are cheap and unreliable, professional telescopes are expensive and also unreliable. Most problems in the operation of a ROTS are usually telescope-related. Development of a common *robotic telescope platform* is much needed to facilitate the implementation of ROTSs.

The data products of a ROTS are turbulence parameters in well-defined physical units, e.g., seeing in arc-seconds or turbulence integral in $m^{1/3}$. Each method could provide such data by relying on its principle, theory, and internal calibrations. External calibration or "validation" is not needed. Of course, it is a good metrology practice to inter-compare the instruments. Inter-comparisons help to detect and eventually eliminate various instrumental biases. An old tradition of *inter-calibrating* seeing monitors, i.e., establishing empirical corrections to bring the data onto a "common system" should be abandoned once and for all. If an instrument produces biased data, empirical corrections do not help because the bias depends on the atmospheric conditions. Two instruments can agree perfectly at one site or on one night (e.g., with slow wind speed) and diverge under different circumstances.

Turbulence measurements differ from most other measurements in one significant aspect: they cannot be repeated. This places a requirement for strict control of the data quality and biases *at the moment of acquisition*. Data quality is a new concept in the field of ROTS; most existing instruments either do not implement it or have only partial quality checks. This will change progressively in the future.

A need for a ROTS is not always justified. Adaptive optics, a major user of turbulence data, can deduce all required turbulence parameters from the AO instrument itself. Such information would be more relevant than the "site data" from a stand-alone ROTS. Development of reliable on-line extraction of atmospheric parameters from the AO loop data is still much needed. Similarly, estimates of the seeing obtained through the telescope are to be preferred over site-monitor seeing for queue-scheduled observatory operation or a posteriori data analysis. Modern telescopes equipped with fast guiders and active optics have all hardware functionality for on-line seeing measurements, but never use it.

Authored by
A. Tokovinin
Cerro Tololo Inter-American Observatory, La Serena, Chile

CHAPTER 4

An Introduction to Measurements with Standard Optical Turbulence Profilers

As a matter of editorial clarification, two things should be noted. First, in choosing the instruments included in this contribution, I was guided by a number of considerations, including the maturity of the techniques in theory and in fielding of the instruments, the historical or prevalent use of a technique for characterizing astronomical sites, and its potential for continued use in the future. As discussed by Tokovinin (Chapter 3) this field is currently undergoing encouraging and rapid growth in the number of techniques and instruments available to study optical turbulence and this contribution ignores, for space, many of the new approaches. It should be noted, however, that campaigns/studies over the past several decades have used a fairly limited set of instruments. These instruments, their basic underlying principles and assumptions, and the data they provide (and do not provide) are the focus of this chapter. Second, my role as first author is largely an editorial one for most of the words within this contribution are from the experts (and in many cases the originators) of the techniques/instruments. The instruments and my co-authors are as follows:

- Differential Image Motion Monitor (DIMM): A. Tokovinin
- Generalized Seeing Monitor (GSM): A. Ziad
- SLOpe Detection And Ranging (SLODAR): R. Wilson
- Generalized SCIntillation Detection And Ranging (G-SCIDAR): R. Avila, M. Chun
- LOw-LAyer SCIDAR (LOLAS): R. Avila
- Multiple-Aperture Scintillation Sensor (MASS): A. Tokovinin
- SOund Detection And Ranging (SODAR): T. Travoullion
- Balloon-based Thermosondes: G. Jumper

The goal of this chapter was to assemble a reference for the major optical turbulence characterization techniques that include their basic principles of operation, the interpretation of the data the technique provides (including their accuracy and precision), and, importantly, the basic use and calibration of the instruments. Given that there are detailed papers on each of these techniques, the intent is not to repeat information but rather to present a summary of what the different techniques can (and cannot) provide and also to offer less rigorous but equally important 'real-life' advice to those trying to use or implement the techniques.

The instruments are discussed in order of the basic data the instrument provides: integrated quantities to turbulence profiles. While the title of this paper refers to instruments of the latter type, the DIMM is so prevalent in site characterization at astronomical sites that it demands to be included. The Generalized Seeing Monitor (GSM), while not as prevalently used, provides a set of integrated parameters and, importantly, provides a means to monitor the wavefront outer-scale. Its use at most of the major telescope sites around the world and in Antarctica (an important future telescope site) amounts to an impressive data resume. The profiling instruments include the "usual suspects" with the novelty of this contribution being the side-by-side comparison.

Throughout the text we use the following nomenclature (Table 4.1). Most of the notation is standard.; see Preface for definitions.

As a final note, many atmospheric parameters depend on wavelength. Within the context of atmospheric characterization of astronomical sites the standard is to specify the parameters at a wavelength of 0.5 microns (500 nm) and corrected to the zenith.

Table 4.1 Summary of techniques

Instrument	Measured quantity	Derived parameters	Inferred parameters
DIMM	Wavefront curvature on single spatial scale (e.g., differential tilt)	r_0, ε_0, total integrated seeing	Existence of clouds, θ_0, τ_0
GSM	Wavefront tilts, scintillation index	Total integrated $C_N^2(h)$, wavefront outer scale, and isoplanatic angle (θ_0)	coherence time (τ_0)
SLODAR	Wavefront tilts	Integrated $C_N^2(h)$ in a fixed number of equally spaced altitude bins, total integrated seeing	Free-atmosphere seeing
GS/LOLAS	Amplitude spatial auto-correlation and cross-correlations	$C_N^2(h)$ at a fixed number of altitudes up to a maximum altitude	GS: moments of the $C_N^2(h)$ profile LOLAS: integrated $C_N^2(h)$ up to maximum altitude.
MASS	Scintillation indices	Integrated $C_N^2(h)$ in a fixed number of slabs from about 250 m to 16 km with a resolution of $\Delta h/h \sim 1$	τ_0 (estimate excluding ground turbulence), isoplanatic angle
SODAR	Backscattering cross section of atmospheric cells at acoustic frequencies	Temperature structure constant $C_T^2(h)$, velocity field of cells	$C_N^2(h)$ if in situ measurements of other meteorological parameters are made.
Thermosondes	Temperature, pressure, humidity	Temperature structure function at a fixed number of spatial scales	$C_N^2(h)$

Table 4.2 Notation

Quantity	Name	Typical Units
$C_N^2(h)$	Index of refraction structure function constant	meters$^{-2/3}$
$\mu_n = \int_0^h C_N^2(h)h^n dh$	Moments of the turbulence profile	meters$^{1/3+n}$
r_0	Fried parameter proportional to $\mu_0^{-3/5}$	meters, cm
ε_0	Seeing resolution	arcseconds
θ_0	Isoplanatic angle proportional to $\mu_{5/3}^{-3/5}$	arcseconds
τ_0	Atmospheric time constant	milliseconds
$\mathcal{L}_0$	Integrated wavefront outer scale	meters

4.1 Differential Image Motion Monitor

4.1.1 Basic Principles

The Differential Image Motion Monitor (DIMM) is an instrument that measures the Fried parameter r_0, or equivalently, the C_N^2 integral over the whole atmosphere, or the seeing $\varepsilon_0 = 0.98 \lambda/r_0$. Usually, a DIMM consists of a small telescope with a 2-aperture mask in front of it (Fig. 4.1). One hole is equipped with a thin-edge prism to produce two separate images of a bright star. This double image is recorded many times by a CCD detector with short (5-10 ms) exposures. Centroids of each spot are calculated and their differential variance (in detector pixels) is then computed over a series of images taken over a period of one minute. The hardware implementation of a DIMM can differ in many ways from the typical

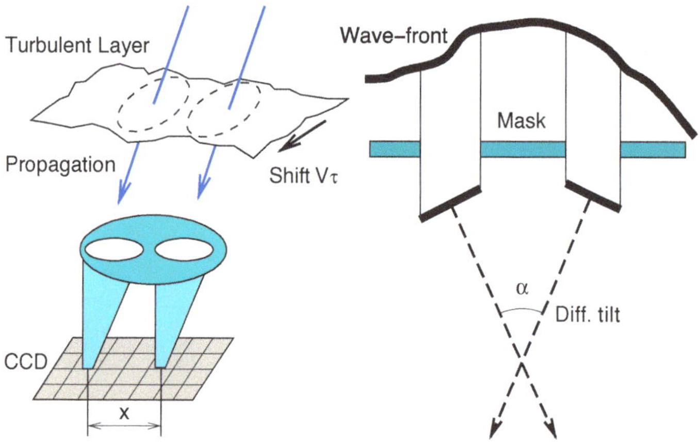

Figure 4.1 The basic principle of a DIMM is illustrated. Two apertures image starlight independently onto a detector to monitor the differential tilt between the two apertures. This is equivalent to measuring the wavefront curvature over a fixed spatial scale.

case described above. Instead of a prism, the spots can be split by mirrors or lenses inside the instrument, as in a MASS-DIMM (Kornilov et al., 2007), or imaged by two separate telescopes. The detectors can be one- or two-dimensional, with or without image intensification. A DIMM can be considered as a 2-aperture Shack-Hartmann wave-front sensor, so this method can be applied to adaptive- or active-optics systems to measure the seeing in a large telescope.

A DIMM is not sensitive to the telescope shake or guiding errors (to the first order) and its field of view is usually wide (few arc-minutes) easing the pointing requirements for the telescope. These features and its relative simplicity contribute to the popularity of this instrument. It is a standard for seeing measurements.

The principal measurement made by the DIMM is the differential wavefront tilt between two apertures separated by a distance. The average separation between the spots is determined by the DIMM optics (prism and telescope focus), but random fluctuations of the separation are caused by the wavefront tilts produced by turbulence at the apertures. Knowing the angular detector pixel size, we can convert the differential centroid fluctuations in pixels to a differential tilt α in radians. To first order, this tilt is related to the overall wave-front curvature over the full baseline of the apertures (Fig. 4.1, right). There are two orthogonal directions over which this differential tilt can be calculated. The tilt in the direction parallel to the holes' centers is called *longitudinal* and the tilt in the orthogonal direction is called *transverse*. The variance of either of these differential tilts σ_α^2 is related to r_0 by

$$\sigma_\alpha^2 = K \, (\lambda/D)^2 \, (D/r_0)^{5/3} \qquad (4.1)$$

Here D is the aperture diameter, the differential wavefront tilt variance σ_α^2 is measured in square radians, and the coefficient K depends on the separation of the holes (baseline) B and the tilt measurement direction (longitudinal or transverse) (see Tokovinin and Kornilov, 2007, for other factors). The calculated r_0 refers to the wavelength λ. It is strongly recommended to use $\lambda = 5 \times 10^{-7}$ m as a standard. The r_0 parameter is often replaced by the *seeing* $\varepsilon_0 = 0.98 \lambda/r_0$ (Roddier, 1981). The two estimates of the seeing (longitudinal and transverse) are generally averaged and corrected for the observation zenith distance (ε_0 is multiplied by the $\cos(\gamma_{zenith})^{3/5}$).

4.1.2 Data and Calibrations

Standard DIMM data include a sequence of seeing ε_0 (at zenith and $\lambda = 500$ nm) as a function of universal time, complemented by additional (non-standard) parameters relevant to the data quality control (zenith angle or air mass, fluxes, average separation between spots, Strehl ratios, scintillation index, etc.).

The above relationship for r_0 and the differential wavefront tilt variance σ_α^2 makes the assumptions that the turbulence is Kolmogorov and that the measurements are made instantaneously (have a zero exposure time). The first assumption is good at spatial scales from D to B relevant to a DIMM while corrections should be made for the latter (see Tokovinin and Kornilov, 2007). The measured variance is always biased by finite exposure time, detector noise, etc. Controlling and correcting biases in a DIMM is an issue treated in detail elsewhere (Martin, 1987; Tokovinin, 2002; Tokovinin and Kornilov, 2007).

For **calibration** of a DIMM, we need to know the aperture diameters (D), their separation (B) ,and the pixel scale. For **de-biasing** the data, we need to know the detector readout noise, some parameters of the spots, and of the centroid algorithm. The exposure-time bias can be either estimated by comparing quasi-simultaneous data with different exposure times (e.g., by binning adjacent images) or reducing it to an acceptable level by using exposures shorter than 5 ms. The most troublesome bias is caused by the interaction between scintillation and imperfect DIMM optics. This bias depends on the usually unknown turbulence profile and aberrations. The only way to avoid this is to maintain a nearly diffraction-limited quality of the spots' images, quantified by, for example, their Strehl ratios. Even with such a control (e.g., both Strehls above 0.6), obtaining absolute accuracy of 10% or better may be difficult (Tokovinin and Kornilov, 2007; Wang et al., 2007). The problem is related to the high-altitude turbulence that produces scintillation and is not well described by the standard weak-perturbation theory, while the low-altitude turbulence is measured more accurately. We may say that the *weighting function* of a DIMM (its response versus altitude) is always one at low altitudes, but becomes somewhat uncertain at higher altitudes because of aberrations and propagation.

In addition to the seeing, a DIMM can provide estimates of other atmospheric parameters. The isoplanatic angle θ_0 can be measured approximately from the flux fluctuations (scintillation index) if $D \sim 0.1$ meters. The atmospheric time constant τ_0 can be estimated if the effective wind speed is known, but it is never actually *measured* by a DIMM. Fluxes in the spots can be used to detect clouds or to determine extinction. None of these extra data products are fundamental to the DIMM method.

Errors in the DIMM implementation or use are most often caused by imperfect optical quality, e.g., by focus changes due to temperature, poor optical alignment, or even intentional defocus. A common error is to confuse the longitudinal direction with the direction parallel to the spots (instead of the baseline). In some DIMMs, the need to correct the exposure-time or noise biases is neglected. An aperture diameter D of at least 0.1 m is recommended; otherwise the diffraction and spectral bandwidth will affect the DIMM weighting function. The detector angular pixel size must be of the order $\lambda/(2D)$ to sample the spots correctly.

The next two instruments discussed make use of the DIMM approach with additional measurements to obtain estimates of the wavefront outer scale and the turbulence profile ($C_N^2(h)$). For each, the DIMM is an integral component of the measurement and as such the calibrations and assumptions discussed above are appropriate for these instruments.

4.2 Generalized Seeing Monitor (GSM)

In its original implementation (call the "Grating Scale Monitor") the GSM was intended to measure the integrated wave front outer scale $\mathcal{L}_0$. Theoretical and experimental developments have extended the technique to also provide measurements of the seeing ε_0, the isoplanatic angle θ_0, and the atmospheric wavefront coherence time τ_0 (Ziad et al., 2000), and the instrument is now referred to as the Generalized Seeing Monitor (GSM). The definition of most of these parameters is given using the Kolmogorov model, which is valid only in the inertial range (spatial range between inner and outer scales). Other models have been proposed for a better understanding of the atmospheric turbulence effects beyond the scale limits. These models are empirical and up to now there was no distinguishing measurement that pointed to when and where a particular model is appropriate. Its most recent incarnation is capable of atmospheric turbulence model verification and the altitude distribution of the wavefront outer scale (e.g., $\mathcal{L}_0(h)$).

The wavefront outer scale is an important parameter for the experimental evaluation of the performance of large aperture telescopes. The actual size of the outer scale has long been controversial, with measured values ranging from less than 10 m to more than 2 km. What is not controversial is the conclusion that when the diameter of the telescope approaches or exceeds the size of the outer scale, the optical consequences of atmospheric turbulence are dramatically changed from their traditional Kolmogorov behavior. In particular, power in the lowest spatial aberration modes, e.g., Zernike tip and tilt, and the overall stroke required for an adaptive-optics system are reduced for a finite outer scale (Winker, 1991). With the current interest in the design of extremely large ground-based optical and infrared telescopes, reliable estimates of the outer scale profile have assumed considerable importance. A finite outer scale has implications for interferometry as well (Mariotti, 1994; Conan, 2000). Indeed, an outer scale smaller than the interferometer baseline leads to a saturation of the fringe excursion when the Kolmogorov model foreshadowed a steady increase with the baseline.

4.2.1 Basic Principles

The first GSM was developed at Nice University and it has been operational since 1997 at La Silla Observatory (Martin et al., 1998). It measures the correlation of Angle of Arrival (AoA) fluctuations at several points of the perturbed wavefront out to spatial scales of about a meter using several small telescopes. Contrary to intuition the optimal scale length for these measurements is smaller than the outer scale. A recent development of the GSM technique is the use of the GSM in the longitudinal or transverse direction with 6 modules instead of 4. The reconstructed AoA covariance combined with some assumptions on the atmospheric turbulence behavior allows the retrieval of the outer scale profile for the prevailing layers. The first results with this method were obtained from a combined GSM-balloons observations at the Cerro Pachon, Chile, in 2000 (Ziad et al., 2004a).

4.2.2 Data and Calibrations

The GSM instrument measures the Angle of Arrival (AoA) fluctuations at different positions of the wavefront. The AoA spatio-temporal correlations are then computed leading to estimates of r_0, outer scale, isoplanatic angle, and coherence time (Ziad et al., 2000).

The original version of the GSM instrument consists of four 10-cm telescopes installed on equatorial mounts and equipped with detection modules measuring the AoA fluctuations (Ziad et al., 2000). Each telescope measures the AoA fluctuations by means of flux modulation which is produced by the displacement of the star image over a Ronchi grating. Two telescopes are installed on a common mount on a central pier (Fig. 4.1) working as a differential image motion monitor (DIMM) with a 25-cm baseline. Two other telescopes have different mounts on separate piers, located 0.8 m to the south and 1 m to the east from the central pier. This configuration has been chosen for more sensitivity to the outer scale.

The GSM version for Dome C (Fig. 4.2) is based on two identical DIMMs observing simultaneously the same star and separated by a distance of 1 m in the NS direction (Ziad et al., 2008). Each DIMM is a telescope equipped with a mask with sub-apertures of diameter 6 cm, distant 20 cm. This mask is placed at the top entrance of the telescope. One of the holes is equipped with a small angle prism (deviation 30 arcsec), the other one with a glass parallel plate.

Figure 4.2 GSM deployed in Antarctica

The AoA covariances are computed for each baseline and normalized by the differential variance of AoA on the 20- or 25-cm baseline. They are compared to von Karman theoretical normalized covariances (Avila et al., 1997b) and the appropriate $\mathcal{L}_0$ is found for each baseline. The final value of $\mathcal{L}_0$ is taken as the median of the individual $\mathcal{L}_0$ values. The seeing ε_0 is deduced from the differential variance as in the DIMM instrument. The scintillation index σ_I^2 is computed during data reduction and used to estimate the isoplanatic angle (Ziad et al., 2000).

A quantification of the different GSM noises has been performed (Ziad et al., 2000) and hence corrections of photon and scintillation noises are done before data processing. Another correction for finite exposure time is also performed; it consists in computing AoA variance (or covariance) for 5 ms and 10 ms and in extrapolating linearly to the 0 ms exposure time. Finally, the statistical errors of the computed variances and covariances are estimated and consequently the errors of the atmospheric optical parameters measured with GSM are provided. Telescope vibration problems are a major uncertainty. To minimize these effects, the quadratic trends are subtracted from the data before calculating covariances, the measurements are made in only the declination direction, and the instrument is protected by wind screens. It is assumed that other telescope-related errors are uncorrelated.

4.3 SLODAR

4.3.1 Basic Principles

SLODAR (Wilson, 2002) is an optical triangulation method for measurement of the atmospheric turbulence profile. Development of SLODAR began with tests at the William Herschel telescope at La Palma (Wilson, 2002). To date, campaigns to characterize the turbulence profile using dedicated SLODAR instruments have been carried out at the Cerro Tololo (Sarazin et al., 2005), Cerro Paranal (Sarazin et al., 2007), Mauna Kea (Chun et al., 2007), and Siding Spring (Goodwin et al., 2007) observatories. The method has also been applied at Mount Palomar, employing the Multi-Guide Star Unit of the AO system at the Palomar 5-m telescope (Wang et al., 2007).

A telescope equipped with a Shack-Hartmann wavefront sensor (WFS) observes a double star, so that two sets of WFS spots (1 for each star) are imaged onto the detector. The turbulence profile is extracted from the spatial cross-covariance of the measured centroids of the WFS spots, typically averaged for a few tens of seconds of data recorded at a frame rate of ~50-200 Hz. The common motion of the spots (the centroid average over all illuminated sub-apertures in the array) is subtracted, to remove the effects of telescope wind-shake or guiding errors. SLODAR can be applied to a small telescope (e.g., 35-50 cm) as a stand-alone site monitor or to a large astronomical telescope—for example making use of WFS measurements from an existing adaptive optical system, if more than one guide star can be observed simultaneously. The temporal evolution of turbulent layers (e.g., translational velocity and velocity dispersion) can also be studied from the spatio-temporal cross-covariance of the WFS measurements. A number of enhancements of the SLODAR method have been proposed, including improvement of the altitude resolution by stepped optical re-conjugation of the lenslet array (Goodwin et al., 2007), and inclusion of the cross-covariance of both centroid and intensity fluctuations of the WFS spots in the data analysis (Vedrenne et al., 2007).

Each turbulent layer at a specific altitude produces a peak in the measured cross-covariance of the WFS centroids. To retrieve the quantitative turbulence profile from the data requires fitting of a theoretical functional form for the tilt cross-covariance as a function of altitude (Butterley et al., 2006). The theoretical covariance, or 'impulse response' of the system to a thin turbulent layer as a function of altitude, can be calculated for any given spatial spectrum of the turbulence. Hence, for example, the turbulence outer scale or the coefficient of the power law for the spatial turbulence power spectrum can be included as variables in the fit to the data. For small sub-apertures ($<\sim 20$ cm) and high-altitude layers, it is important to take into account the effects of Fresnel propagation (scintillation) on the tilt covariance function (Goodwin et al., 2007).

4.3.2 Data and Calibrations

SLODAR provides measures of the integrated turbulence strength in a number of bins of equal size (H) in range along the line of sight, with the first (lowest altitude) bin centered at the ground (corresponding to zero spatial offset in the cross-covariance). The maximum number of resolution elements (N) is equal to the number of WFS sub-apertures subtended across the telescope aperture, so that the maximum altitude for profiling is given by (N-1)×H. The total integrated turbulence strength for the entire path through the atmosphere can also be measured from the auto-covariance function of the centroids, or by a differential image motion analysis for pairs of sub-apertures. Hence the "unsensed" contribution (total turbulence strength for all altitudes above (N-1)×H can be found by differencing the total turbulence strength and the integral of the directly measured profile.

Figure 4.3 is an example of the output of a SLODAR profiler for a six-hour period on 1 December 2004, during a campaign to investigate the ground-layer turbulence at the Cerro Tololo observatory (Sarazin et al., 2005). Here we see traces representing the turbulence in the eight directly sampled bins up to a maximum altitude of 1200 m, and also the higher altitude total (top). Turbulence strength is represented by the width of each trace. The variation in resolution of the profiler with the changing altitude of the target star can be seen. In normal operation, the spot centroids and other calculations can be carried continuously so that the updated turbulence profile is available within a few seconds of 'real time'.

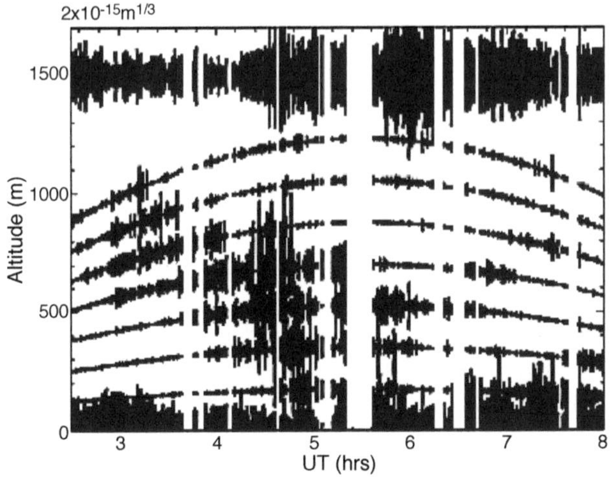

Figure 4.3 Example SLODAR turbulence profile for the night of 1 December 2004 at Cerro Tololo Observatory in Chile. Trace widths indicate the strength of the turbulence in the sampling bins with a scale shown at the upper right of the figure and the upper bars denote the integrated strength in the free-atmosphere.

The measured centroid cross-covariance is not biased by shot noise (we have different photons from each of the two stars). Hence relatively faint target stars can be used while still permitting accurate profile measurements. However, for faint targets, the signal-to-noise ratio for each instantaneous WFS measurement may be low. The overall sampling time required to measure the strength of a given layer depends on the SNR per WFS sample and the rate at which independent samples are acquired (note that this rate is lower for slow turbulence wind speeds). In typical conditions, the integrated turbulence strength in each altitude bin can be determined with an rms uncertainty of better than 1×10^{-14} m$^{1/3}$ in less than one minute of continuous sampling.

A fundamental limitation of the SLODAR technique is that a bright double-star target is required. For dedicated SLODAR profilers based on small telescopes, and hence with small sub-apertures (5-10 cm), there are typically only a small number of bright double stars with the required angular separations. For such systems, a WFS detector with high quantum efficiency and very low effective read-out noise—for example a CCD camera equipped with electron multiplication technology or an image intensifier—is required to access sufficient targets for continuous observations (e.g., at least one target visible always above 45° elevation at all times).

The altitude resolution and maximum altitude for profiling are limited by the number of sub-apertures across the WFS. For a given system, observing a wider double star can increase the resolution, but the max altitude will be proportionally reduced. Hence for a small telescope we may make higher-resolution measurements of the ground-layer (wide target) or lower-resolution measurements of the full profile (narrow target) but not both at the same time. The achievable altitude resolution is ultimately limited by the effects of diffraction, photon noise, and scintillation, imposing a minimum sub-aperture diameter of ~5 cm. The maximum double star separation is determined only by the optical parameters of the telescope and WFS.

Accurate calibration of the angular image scale at the WFS detector is critical for SLODAR, since the turbulence strength is given by the (co)variance of the angular centroid motion. This can be achieved by observing double stars of well known separation. It is important to place the lenslet array accurately at the optical conjugate of the telescope entrance aperture, so that the mapping of the telescope pupil onto the array is identical for the two stars of the target double. An error in the lenslet position will result in an error in the zero point of the altitude scale—for example a relative misregistration of the re-imaged pupils for the two stars equivalent to one tenth of a sub-aperture implies an altitude zero-point error of one tenth of the profiling bin size.

Although SLODAR can be implemented effectively on small telescopes, good optical and mechanical stability is required. If there are large image motions (e.g., more than few arc-seconds peak-to-peak) due to windshake it is difficult to track centroid motions accurately while maintaining real-time processing at high frame rates. Furthermore, the WFS frame rate will be reduced if a large detector region-of-interest is required to accommodate large image excursions due to wind-shake. For the same reason, auto-guiding of the telescope may be required to minimize telescope tracking errors. If necessary, auto-guiding offsets can be measured directly from the WFS images.

Optical distortions due to changes of the telescope collimation and focus (e.g., with elevation and temperature) can be problematic for commercial amateur telescopes. For many such telescopes (e.g., Meade, Celestron), focus is achieved by movement of the primary mirror. This may result in sudden, large changes of the image position, focus, and/or collimation. For these telescopes the primary mirror must be effectively locked (often requiring additional custom locking mechanisms) and focusing of the WFS achieved via an accurate focusing mechanism, preferably with powered remote control via computer.

Turbulence above the telescope primary mirror can be strong, particularly for small telescopes with a closed-tube design, and an open-truss optical tube assembly is preferable. This local turbulence appears in the lowest SLODAR resolution element, effectively contaminating the measurement of the 'true' surface layer. The mirror or tube turbulence typically appears in the power spectra of the raw centroid data at frequencies much lower than those expected for true atmospheric layers (which have a well-defined wind velocity). Hence the mirror/tube seeing can be removed from the data by application of a low-pass temporal filter (Goodwin et al., 2007). However, care must be taken with the cut-off frequency of the filter, particularly when the surface-level wind speed is low.

4.4 G-SCIDAR/LOLAS

4.4.1 Basic Principles

For many years the standard of optical turbulence profiling techniques has been Scintillation Detection and Ranging (SCIDAR) or its more common form Generalized-SCIDAR. The technique relies on the correlation of amplitude variations within the complex optical wavefront from a pair of double stars. The original SCIDAR implementation correlated amplitude variations seen in pupil plane images. Since phase perturbations of the wavefront must propagate a distance before they turn into amplitude variations, the classical SCIDAR implementation cannot measure wavefront phase perturbations arising near the pupil plane (this is typically at or near the ground). The Generalized SCIDAR (GS) extension of the technique optically conjugates the detection plane to a couple of kilometers below the ground. This virtual path length allows the phase perturbations arising near the ground to propagate to amplitude variations (Avila et al., 1997a; Fuchs et al., 1998). This technique is generally done on moderately large telescopes (1.5-2 meters) in order to obtain sufficient Signal to Noise Ratio (SNR) in individual data sets of a few seconds.

Recently, Avila et al. (2008) have applied the GS technique to widely separated double stars using an instrument outfitted to a small portable telescope. This technique is featured here, though much of the discussion is applicable to the GS technique. The new instrument, LOw-LAyer SCIDAR (LOLAS), aims to measure the C_N^2 (h) profile in the atmospheric boundary layer with high-vertical resolution. Such measurements are a necessity for the development of ground-layer adaptive optics (Rigaut, 2002; Tokovinin, 2004) systems, as pointed out by, e.g., LeLouarn and Hubin (2006), and they are also crucial for the study of the performances of next-generation giant telescopes. Avila et al. (2008) give a more thorough description of the instrument.

The GS and LOLAS techniques have the same theoretical basis and they are presented here as one and the same. The actual implementations differ but we will delay that discussion to a later section. The GS concept (Avila et al., 1997a; Fuchs et al., 1998) consists of computing the normalized mean spatial autocorrelation function of short exposure-time images of the scintillation pattern produced by a double star on a detector which is made the conjugate of a virtual plane (analysis plane) at a distance h_{gs}, of the order of a few kilometers, below the telescope pupil (e.g., $h_{gs}<0$). Hereafter, within this section, we will refer to this normalized-mean autocorrelation simply as autocorrelation. Following Avila et al. (1997a) and references therein, the autocorrelation function obtained can be written as

$$C_{g_sh_{gs}}(\mathbf{r}) = \int^{+\infty} dh\, K\,(\mathbf{r}, h - h_{gs}, \mathbf{\rho})\, C_N^2\,(h) + N(\mathbf{r}), \tag{4.2}$$

where the kernel $K\,(\mathbf{r}, h - h_{gs}, \mathbf{\rho})$ is the theoretical autocorrelation function produced by a single layer at an altitude h with a unit C_N^2, and $N(\mathbf{r})$ is the estimated noise. When $h<0$, $C_N^2 = 0$. For a double star of angular separation $\mathbf{\rho}$, the kernel consists of three autocorrelation peaks: one centered on the autocorrelation origin and the other two separated from the first one by $+\mathbf{\rho}|h-h_{gs}|$ and $-\mathbf{\rho}|h-h_{gs}|$.

The altitude h of the layer is determined by the position of either of those lateral peaks, because $\mathbf{\rho}$ and h_{gs} are known parameters. The accuracy of the altitude is determined by the accuracy of this position determination. If the correlation peaks are properly sampled spatially by the detector (e.g., two pixels), then the resolution is approximately equal to one pixel. Let d_p be the pixel size projected on the pupil. This is also the pixel size in the autocorrelation plane. The altitude resolution of the retrieved $C_N^2(h)$ profiles is given by $\Delta h = d_p/\rho$ where ρ is the modulus of $\mathbf{\rho}$.

The maximum altitude, h_{max}, for which the C_N^2 value can be retrieved is set by the altitude at which the projections of the pupil along the direction of each star cease to overlap, as no correlated speckles would lie on the scintillation images coming from each star. The maximum altitude is thus given by $h_{max} = D/\rho$, where D is the pupil diameter.

In addition to the autocorrelation of the scintillation images, since the GS calculates the mean cross-correlation of images taken at times separated by a known constant time delay Δt, the temporal cross-correlation leads to the determination of the velocity of the turbulent layers (Avila et al., 2006; Prieur et al., 2004; Avila et al., 2006). This is particularly useful for distinguishing the optical turbulence generated that arises from inside the telescope dome or tube.

4.4.2 Data and Calibrations

The LOLAS concept consists of the implementation of the GS technique on a 40-cm dedicated telescope, using a very widely separated double star. For example, for $d_p=1$ cm and star separations of 180″ and 70″, Δh equals 11.5 and 29.5 m, while h_{max} equals 458 and 1179 m, respectively. The value of d_p is set by the condition that the smallest scintillation speckles be sampled at the Nyquist spatial frequency or better. The full width at half maximum of the autocorrelation of the scintillation produced by a phase screen at a distance L from the detection plane is $\Delta R = 0.78\sqrt{\lambda L}$ (Prieur et al., 2001), where λ is the wavelength. This corresponds to the typical size of scintillation speckles produced by that phase screen. The shortest value of L corresponds to $|h_{gs}|$. A value of -2 km is chosen for h_{gs}. This value is justified in the discussion below. Taking $L=|h_{gs}|$ and $\lambda = 0.5\mu m$, the value for ΔR is 2.45 cm. Thus the value of 1 cm for the spatial sampling spacing d_p satisfies the Nyquist condition, namely $d_p<\Delta R/2$.

Figure 4.4 (top) Photograph of the LOLAS instrument on the Coude roof of the UH 2.2-m telescope on Mauna Kea, Hawaii. (bottom) Plot shows a CN2 profile obtained using a nearly 200 arcsecond-separation double-star. The vertical resolution for that profile is 8 meters. The minimum detectable C_N^2 value is 5×10^{-16} m$^{-2/3}$. Note the ability of the profiler to discern a layer centered at 16 m above the ground layer. The turbulence generated inside the telescope tube has been removed.

The propagation distance from the pupil to the analysis plane, h_{gs}, is set to -2 km, as a result of a compromise between the increase of scintillation variance, which is proportional to $h_{gs}^{5/6}$, and the reduction of pupil diffraction effects. Indeed, pupil diffraction caused by not being at the pupil plane means that Eq. 4.2 is only an approximation. The larger the $|h_{gs}|$ or the smaller the pupil diameter, the greater the error in this approximation. Numerical simulations show that the error introduced by the pupil diffraction effect in the autocorrelation values is less than 10% for a 40-cm aperture telescope and an h_{gs} = -2 km conjugate layer. This discrepancy is corrected for in the data reduction procedures.

The out-of-focus pupil images produced by each star are centered on the detector. To minimize the acquisition and processing speed, only a subframe of 256×80 binned pixels is acquired. The exposure time of each frame is selected by the user to be between 3 to 10 ms, depending on the wind conditions and photon flux from the stars. The typical number of frames used to obtain one set of auto- and cross-correlations is 30000, which corresponds to about 5 minutes of data.

In windy conditions, telescope shaking can cause image motion which affects the calculation of the mean image. To avoid this, every image is re-centered prior to summing the images. In addition, eventual tracking drifts are corrected by updating the telescope position every 200 images, using the average image position of the most recently acquired 200-frame packet. This auto-guiding assures that the stars remain in the active portion of the detector. It has been seen that the pupil images slowly change in size. This is due to a slow shift of the telescope focus, presumably caused by the redistribution of the primary mirror load while tracking the stars. An auto-focus system has been developed to overcome this problem. On every average image calculated with 200 frames, the size of the out-of-focus pupil is monitored and the camera and optics are moved by acting on the supporting sliding stage, until the nominal size of 40 binned pixels is recovered. Similarly, the alignment of the stars along the pixel lines is measured and eventually corrected automatically by acting on the motorized rotator. Finally, the average flux over the pupil is also monitored in each image and its variance relative to the mean flux is calculated and stored in the header of the Flexible Image Transport System (FITS) file in which the correlations are saved. This information is used to qualify the data. Indeed, mean flux variations, caused by cloud passages or fog condensation, can adversely affect the retrieval of C_N^2 profiles from the autocorrelations.

Here we list what we consider the most important aspects that have to be taken into account when observing with LOLAS or developing an instrument alike.

- **Targets**: Although suitable double stars for LOLAS measurements are observable year-round and at any location on Earth, those that lead to an altitude resolution better than 12 m are not available from April 8 to 22 and May 22 to June 7 (approximately) at latitudes lower than -13 degrees. This is primarily limited by the required photon flux and requirement for two stars at particular separations.

- **Calibration of h_{gs}**: The optical system is designed to place the detector plane at a particular virtual distance from the pupil plane. Nominally h_{gs} = -2 km. This value must be verified since placement and manufacturing tolerances of the optical elements can change the conjugation plane. The detector conjugate plane must be measured to a precision of the order of 0.6% for a precision of the altitude origin of the order of 10 meters. One way of doing this is to measure the separation of the two extra-pupil images from a double star with a very well known separation (e.g., position errors of 40 milliseconds or less). In long exposure images, the separation of the two extra pupils images can be measured to a 0.1 pixel precision.

- **Telescope and mount**: In addition to the robustness of the instrument, a quality telescope and telescope mount reduces the focus instability, vibrations caused by tracking and wind, and tracking errors.

4.5 MASS

4.5.1 Basic Principles

The MASS instrument is based on the analysis of stellar scintillation. The light from a single bright star is received by a system of 4 concentric apertures— the inner aperture A of 2 cm diameter, and the annular apertures B, C, and D surrounding it, with the outer diameter of D about 8 cm. Photon counts in each aperture are detected with photo-multipliers with 1 ms exposure time. Series of these counts are processed statistically to determine a low-resolution turbulence profile. The apertures and their combinations act as spatial filters, isolating details of certain size. The characteristic size of scintillation pattern is of the order of Fresnel radius $\sqrt{\lambda z}$ ~10 cm for a 10-km propagation (where λ is the wavelength of light and z is the propagation distance). We distinguish contributions from different altitude s by spatial filtering.

A MASS instrument is attached to a feeding telescope with a clear (unvignetted) aperture of at least 8 cm. The beam splitting between annular apertures is done by internal optics (Kornilov et al., 2003). Most MASS instruments in use now are combined with a DIMM and utilize an off-axis part of telescope's aperture (Kornilov et al., 2007). MASS needs only a small telescope; it is insensitive to pointing errors and well suited for turbulence monitoring using single stars of V=2.5 *mag* or brighter. MASS does not sense the near-ground turbulence and does not work under strong scintillation.

4.5.2 Data and Calibrations

After each 1-min. accumulation time, the MASS software calculates 10 *scintillation* indices (4 normal and 6 differential) (Tokovinin et al., 2003; Kornilov, 2007). The relation between turbulence strength, propagation distance z, and index s^2, is a *weighting function* (see Fig. 4.5): $s^2 = \int W(z) C_N^2 dz$. The data are modeled by 6 thin turbulent layers at fixed altitudes h_i of 0.5, 1, 2, 4, 8, and 16 km. The MASS data products are 6 turbulence integrals $J_i = C_N^2(h_i) dh$ in these layers. To compare these numbers with a real, continuous $C_N^2(h)$ profile, we assume that the turbulence integrals J_i correspond to triangular response functions centered on their respective layers. The altitude resolution of MASS is $\Delta h/h \sim 1$. The MASS software automatically accounts for the air mass, spectral type of the star, finite exposure time, and photon noise. Deviations from the weak-perturbation

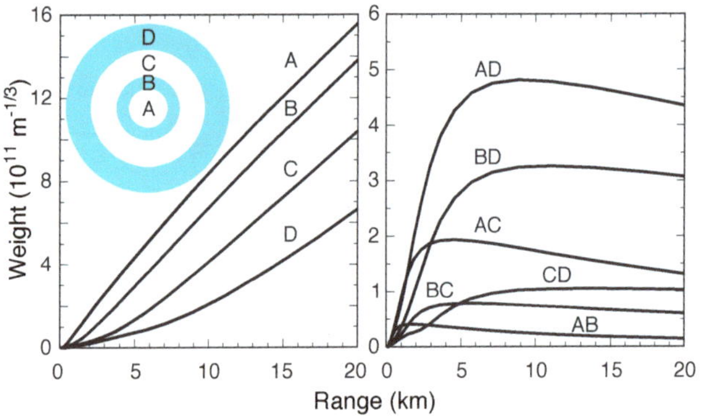

Figure 4.5 MASS apertures and weighting functions.

theory are also corrected up to a certain limit (Tokovinin and Kornilov, 2007).

The atmospheric time constant τ_0 is estimated from the temporal bandpass-filtered scintillation in the smallest aperture A. This method is only approximate and does not include the ground-layer turbulence. Secondary data products of MASS are the free-atmosphere seeing calculated from the sum of J_i and the isoplanatic angle θ_0. The 1-min. average flux can be used for cloud detection and extinction measurement.

Calibration of MASS consists in determining the optical magnification factor, hence the size of the apertures projected on the pupil. The noise parameters of the detectors are determined with an internal light source inside the instrument. The spectral sensitivity curve must be known for correct calculation of $W(z)$ (Kornilov et al., 2007). The sensitivity of MASS data products to the errors in calibration parameters has been studied by Els et al. (2008). They find that the error in the free-atmosphere seeing is less than 0.05″, the integrals J_i are measured to within $\pm 10^{-14} m^{1/3}$ (with a smaller error for high layers). In the model-fitting processes, the integrated atmospheric parameters such as seeing are well constrained, but the individual J_i may be re-distributed between adjacent layers under the influence of random errors and small calibration errors.

Common mistakes in the use of MASS are pointing stars fainter than V=2.5 magnitudes, incorrect setting of calibration parameters or time in the computer, vignetting of the apertures caused by optical misalignment. The data quality parameters recorded in the MASS files can help in detecting some (but not all) of those errors. Regular detector checks with an internal light source are strongly recommended.

4.6 SODAR

SODARs have been very popular instrument with meteorologists and boundary layer physicists. A thorough report on the instrument can be found in Neff (1975).

It is only recently that it has been more widely accepted for astronomical site testing. Early adopters include Erasmus and Banes (1989) for the study of Mauna Kea, and Bedulin et al. (1991) at Maydanak. Antarctic sites such as South Pole and Dome C (Travouillon et al., 2006; Lawrence et al., 2004) have also been studied with SODAR and are probably the first SODAR results of astronomical nature that have been confirmed by independent methods. More recently the TMT site testing campaign has included SODARs in its suite of instruments to qualify the boundary layer conditions of its candidate sites. The results, which will be published at a later date, will be the first large dataset of inter-comparable SODAR measurements.

4.6.1 Basic Principles

The SODAR (SOund Detection And Ranging) is one of few acoustic-based instruments used in astronomy site testing. This is possible because the refraction of light as it travels through our atmosphere mainly depends on the temperature structure constant C_T^2 which in turn is converted to the refractive index structure constant C_N^2 using:

$$C_N^2 = (79. \times 10^{-6} P/T^2)^2 \, C_T^2 \qquad (4.3)$$

where P and T are respectively the atmospheric pressure (mB) and the temperature (K). SODARs measure the temperature structure profile of the lower atmosphere by emitting a series of highly directional acoustic pulses. As the sound travels up in the atmosphere, pressure waves get backscattered by turbulence cells. The backscattering cross-section of such cells is proportional to C_T^2:

$$\sigma^2 = \frac{0.0039 k^{-1/3} \, C_T^2}{T^2} \qquad (4.4)$$

Where k is the acoustic wave number of the emitted pulse. SODARs measure the backscattering cross-section by measuring the ratio of returned signal from a given distant calculated from the pulse delay. Further technical information can be found in Brown and Hall (1978).

In addition to the turbulence which is calculated from the strength of the backscatter, the velocity of the cell can be derived from the Doppler shift of the acoustic pulse.

4.6.2 Data and Calibrations

No commercial unit comes pre-calibrated (see calibration section), so the SODAR outputs a profile of backscatter strength as a function of altitude. Typical range and resolution vary and depend on the SODAR's output power and pulse characteristic.

Units usually offer a range from 200-2000 m and a resolution between 5-30 m. This range and resolution are perfectly adequate to determine the characteristics of the boundary layer conditions at an astronomical site.

In order to achieve a necessary SNR, the SODAR must integrate over a period ranging from several minutes to half an hour. This integration time depends on the desired range and the environmental conditions such as acoustic and electrical noise of the location where the antenna is located.

Over the same range and resolution the SODAR also derives the wind speed and direction. This is possible by emitting pulses in three different directions. Modern SODARs are monostatic, meaning that the point of emission is also the point of reception of the acoustic pulses. They are composed of an array of transducers that can emit sound pulses vertically or, by phasing its emission, at the desired angle to obtain each wind speed component from the relative Doppler shift of the beam. Thanks to this combination of turbulence and wind speed profiling, the SODAR can calculate both the seeing and the coherence time components of the boundary layer.

Calibration has always been a drawback of SODARs for use in astronomical site testing. The difficulty of calculating C_T^2 from the backscatter strength comes from two factors. First, the instrumental characteristic of the instruments, such as gain and noise, must be properly known and constant. These issues have been properly addressed by Danilov et al. (1992); however, their approach does not correct for the second factor. This factor is the humidity and temperature dependence of sound absorption by air and is therefore variable in time (see Harris, 1966). This effect is far from negligible, as experimentally demonstrated by Haugen and Kaimal (1978). Another calibration method was therefore introduced by Travouillon (2006) that includes direct measurements of meteorological parameters to calibrate simultaneous SODAR profiles. This method, which also bypasses the required knowledge of the instrument electronic characteristics, is therefore usable on commercially available SODARs where such parameters maybe difficult to obtain. This calibration has so far produced turbulence measurements in agreement with other methods, such as ground-layer seeing values produced by a DIMM-MASS instrument, within an accuracy of 20%.

Experience has shown that the acoustic quality of the site has more to do with the quality of the measurements than any other parameters. Locations with poor electrical grounding are difficult for the SODAR. Sources of white acoustic noise, such as wind blowing in trees are, even more difficult and can make SODARs totally unusable in such locations. The bare sites of Chile are therefore good for SODARs because of their total lack of vegetation. Antarctic sites work even better, because—besides their low acoustic noise—the temperature range there affects the variation of air absorption far less than do temperate sites. In order to maintain quality control, it is important to maintain close control on noise sources and keep the SNR to an acceptable level.

SODARs are relatively easy to maintain due to their lack of moving parts. They are built from long-lasting components, but transducers may need a regular check for calibration purposes. The main enemies of SODARs in remote sites are lightning and electrical discharge through the ground. Special care must therefore be undertaken to electrically isolate them before deployment.

4.7 Thermosondes

The thermosonde (Titterton et al., 1971; Bufton, 1975) has been used in optical turbulence research since 1971. It is attached to a meteorological radiosonde and is carried into the atmosphere by a balloon. The thermosondes flown by Air Force Research Laboratory (AFRL) personnel use two very fine resistance wire probes to measure the temperature difference across a 1-m horizontal distance. Onboard electronics convert the temperature difference to a voltage signal, amplify and filter the signal, then perform a running root mean square (rms) average over a 4- to 8-second period. The signal is then transmitted to the ground station along with the standard meteorological data: temperature, pressure, humidity, wind speed and direction sensed by the radiosonde (Brown et al., 1982; Jumper et al., 1997). The University of Nice, France, has extensive thermosonde experience as well. Their instrument has a common origin to the U.S. instruments, but which evolved separately from an early Italian design (Barletti et al. 1977). The French thermosondes use 4-wire sensors to measure two separate structure functions with lengths of 0.3 m and 0.95 m (Azouit and Vernin, 2005).

4.7.1 Basic Principles of Operations

The use of the thermosonde to estimate atmospheric turbulence is based on the statistical theory of a randomly varying media. If two temperature sensors pass through the atmosphere at distance, r, apart along a path, x, the mean value of the square of the difference is defined as the temperature structure function, $D_T(r)$ (in units of K^2) (Beland 1993):

$$D_T(\mathbf{r}) = \langle [T(x) - T(x + r)]^2 \rangle \qquad (4.5)$$

Assuming that the temperature is a passive and conservative scalar, and that the dynamic atmosphere follows Kolmogorov (1991) behavior, Obukhov (1949) and Yaglom (1949) deduced that the structure function of the temperature field is expressed in terms of temperature structure constant, C_T^2 ($K^2 m^{-2/3}$):

$$D_T(r) = C_T^2 r^{2/3} \qquad (4.6)$$

The equation applies for distances less than the largest eddies known as the outer scale, L_0, which are on the scale of the turbulent kinetic energy source, and greater than the smallest eddies known as the inner scale, l_0, which are the scales of molecular energy dissipation.

Propagation disturbances are functions of the structure constant of the index of refraction of the air, C_N^2 (m$^{-2/3}$). Index of refraction is related to air density through the Gladstone Dale Constant. Turbulence in the free atmosphere is slow enough that variations in density are essentially due to the turbulent fluctuations of temperature, since pressure disturbances dissipate at the speed of sound. For microwave radiation, the variations in humidity must also be considered. Conversion of the P temperature structure constant to the refractive index structure constant, C_N^2, depends on local pressure, P, mean temperature, T, and the wavelength, λ, of the radiation that is being propagated. For radiation near the visible spectrum, and when the moisture contribution can safely be ignored (normal except right over a body of water), it is customary to use the Eq. 4.3 above.

The theory described above implies a long run through a stationary, randomly varying temperature field in an equilibrium cascade. For the thermosonde, the conditions are varying as we ascend at anywhere from 3 to 8 m/s through what may or may not be stationary, homogeneous, equilibrium conditions. Therefore, we refer to our results as an estimate of the structure constants.

The actual value of C_N^2 at any altitude is not that important, since the impact of the turbulence on optical performance is the integrated effect through the C_N^2 field. For instance, given a C_N^2 field, the path integral equation for seeing (arcsec) is

$$\varepsilon_0 = 5.25\lambda^{-1/5}\left[\int_z C_N^2\, dz\right]^{3/5} \qquad (4.7)$$

The AFRL thermosondes have temperature sensors 1 meter apart, as shown in Fig. 4.6, measuring a one-meter structure function. Measurement across 1 meter simplifies the process since the structure constant is equal to the 1-m structure function. The resistance wire for this instrument is tungsten with a diameter of 3.45 μm. The length of the U.S. wire sensor is less than 5 mm, stretched between two conducting pins, as shown in Fig. 4.7. The nominal resistance is 27Ω, and the temperature coefficient of resistivity is 0.00375 K^{-1}. The current through the wires is approximately 150μA. The wires are approximately perpendicular to the airflow as the balloon lifts the instrument through the atmosphere; similar to typical "hot-wire" and "cold-wire" probes in wind tunnels. U.S. sondes are typically launched with a target ascent speed of 5 to 7 m/s, which requires an initial balloon diameter of around 1.75m using a 1.2 kg balloon, lifting a 2.73 kg payload.

The pair of sensor wires form two legs of a Wheatstone Bridge that generates a rapidly varying voltage based on the difference in the resistances of the wires. The signals are amplified, filtered, and then averaged. The U.S. design uses a circuit that performs a running average with a time-constant of 4 to 8 seconds.

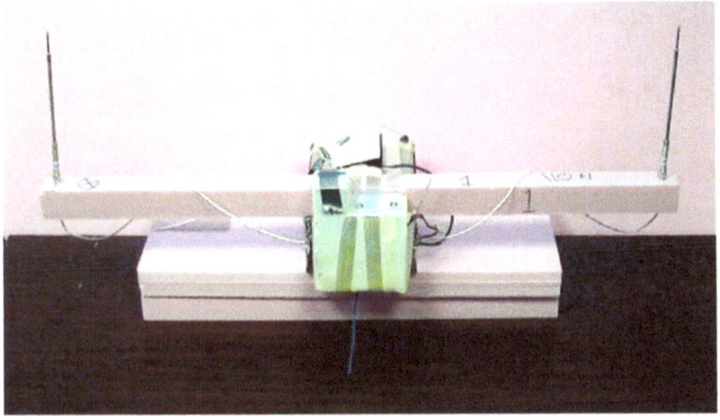

Figure 4.6 A U.S. thermosonde with a 1-m distance between the two sensors on the boom. Radiosonde is taped in front, structure function electronics, batteries, and spare channel interface are behind.

Figure 4.7 Sensor wire-end of U.S. probe; resistance wire is strung between the two electrodes. (below)

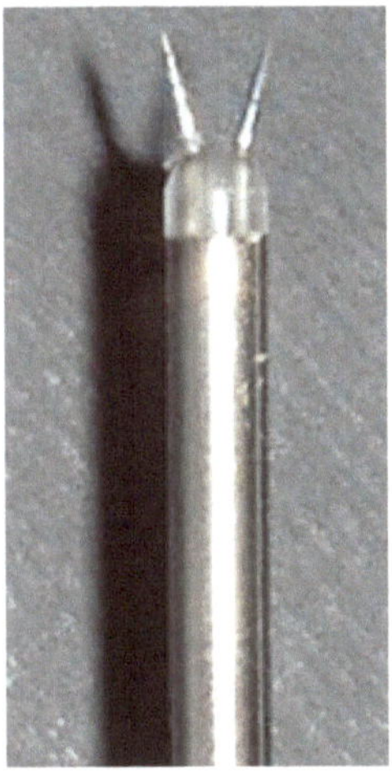

Vaisala radiosondes with spare channel capability are used to send the running average of the structure function and some ancillary information down to the ground station with the typical atmospheric parameters sensed by weather balloon instruments: ambient temperature, pressure, relative humidity, and wind speed and direction. Altitude is determined by the hydrostatic equation, corrected for humidity. Wind instrumentation is determined by GPS, assuming that the horizontal velocity of the balloon is the wind velocity. A comparison of the US and French thermosonde results, along with the Generalized Sidar (GS) can be found in Jumper et al. (2005). There have also been numerous comparisons to other optical turbulence measurement instruments, for instance (Roadcap and Murphy, 1999).

4.7.2 Data and Calibrations

A composite plot of typical thermosonde products is shown in Fig. 4.8. The radiosonde provides the temperature and humidity (shown in the left pane) and the wind speed and direction (shown in the right pane). The temperature and pressure are used with the C_T^2 measurements to compute C_N^2. The results are shown as profiles, but it must be remembered that the balloon is drifting along with the wind, and can travel hundreds of kilometers as it climbs up to an altitude of 30 km, if the balloon does not break prematurely. Also shown in Fig. 4.8 is the climatological C_N^2

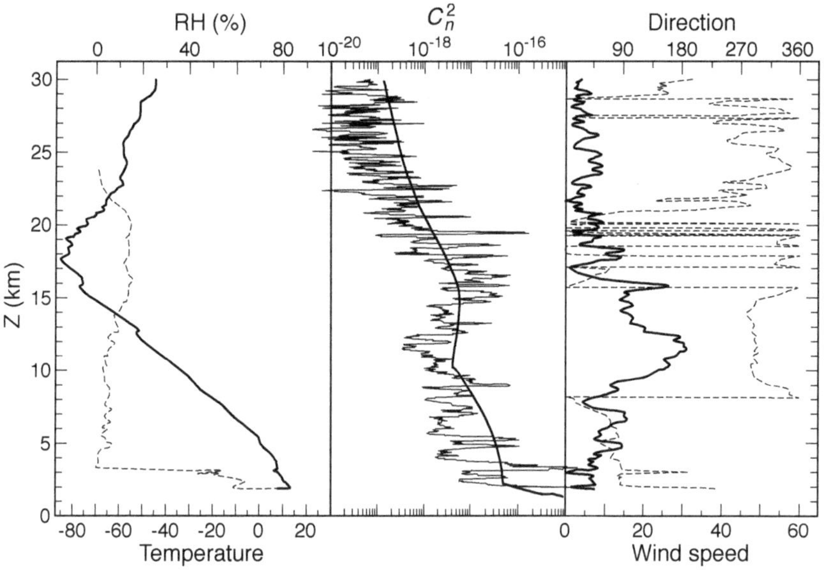

Figure 4.8 An example of the products from a thermosonde: temperature (°C) and humidity, in the left pane, C_N^2 is in the center pane, and wind speed and direction in the right pane. The smooth line in the C_N^2 plot is the climatological model CLEAR1.

model CLEAR1 (Beland 1993). The agreement of French and US thermosondes is generally good, and compares well with GS results, especially at the lower altitudes before the thermosonde has drifted too far from the GS viewing arc, as discussed in Jumper et al. (2005). The C_N^2 profile shows a large variance over as high as 3 orders of magnitude within a narrow altitude band, but generally C_N^2 is decreasing from $\sim 10^{-15}$ m$^{-(2/3)}$ at low altitudes to $\sim 10^{-19}$ m$^{-(2/3)}$ at 30 km primarily due to the effect of pressure in Eq. 4.3. The variance reflects the variations in C_T^2 from a noise floor of about 10^{-6} K m$^{-(2/3)}$ up to a maximum around 10^{-2} K m$^{-(2/3)}$, with most activity around 10^{-4}. The magnitude of C_T^2 increases in the stratosphere where the temperature gradient increases. Composite graphs such as Fig. 4.3 are useful as one attempts to correlate high C_N^2 regions with atmospheric features such as the tropopause and regions of higher wind shear.

The high spatial variations with altitude make the profile difficult to interpret. Many authors address this problem by binning the values to altitude ranges. Another useful display is to show the degradation of an integrated parameter such as seeing, starting at the top with essentially perfect seeing, then showing the increase in beam spread as the light propagates through the atmos-phere (McHugh et al., 2008), as shown in Fig. 4.9.

The thermosonde results are subject to the usual accuracy and precision con-cerns of other instruments. The C_N^2 values depend on C_T^2 from the thermosonde and

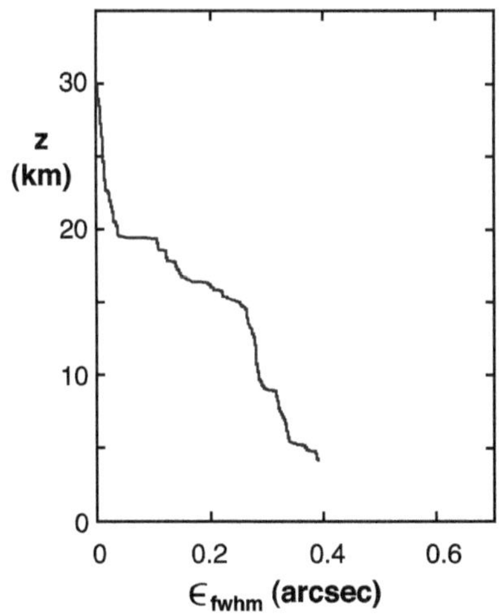

30

z
(km)

20

10

0

0   0.2   0.4   0.6

ϵ_{fwhm} **(arcsec)**

Figure 4.9 Profile of the degradation of seeing, from perfect at the maximum balloon altitude to the value near the ground.

on temperature and pressure measured by the radiosonde (http://www.vaisala.com/weather/products/sounding equipment/radiosondes/rs80). Generally the radiosonde results will be most accurate at sea level, losing accuracy with altitude. The C_T^2 is most accurate at high values, and less accurate at low values near the electronic noise floor. The overall uncertainty in C_N^2 is less than 0.5% at low altitudes, but noticeably increases especially above 20 km, reaching a 4% uncertainty at 30km, mainly due to pressure uncertainties (Brown and Beland, 1988, Murphy et al., 2007).

There are some other issues that should be kept in mind. One problem is that the thermosondes have shown unreliable results in the presence of sunlight, which is not much of a problem in astronomy. There are also heat transfer problems that degrade accuracy at altitudes above 30 km. The thermosonde methodology is based on Kolmogorov's hypothesis, which does not always apply. Other experimenters have used balloon-borne, multiple high-bandwidth temperature sensors to investigate this problem (Otten et al., 2000; Eaton et al., 2009). The French thermosondes, with dual structure functions, provide some insight into the Kolmogorov problem (see reference). Another problem is the contamination of the C_N^2 results by encounters with the wake of the ascending balloon. The high-bandwidth experiments mentioned above were designed to avoid wake encounters. AFRL uses a line in excess of 100 m in length to help avoid the wake. Wake encounters occur when the wind shear is low, which is usually when the measured C_N^2 is low. An investigation with some sensitive temperture probes to locate possible wake encouters showed that regions of high C_N^2 did not have wake signatures, but the low C_N^2 regions did show wake encounters, and probably had even lower turbulence than measured due to the wake (Jumper et al., 2002). It was also determined that the standard RMS chip used in the onboard averaging circuit has a time constant that is slower on decreasing values of D_T^2 than on increasing values. For instance, the AFRL thermosonde uses 4 seconds for increasing values, which results in 8 seconds for the time constant with decreasing values. This problem results in elevated values of integrated C_N^2; this then results in a maximum increase of a factor of 2 for extremely short "impulse" spikes, but it quickly drops down to negligible increases for the thick layers that cause the biggest problems (Jumper et al., 1999). To sum up these problems, if the thermosonde is reading high C_N^2 then the values are probably accurate; if the reading is low C_N^2, the readings may not be as accurate.*

Some other factors to consider are that ascents through rain can break the sensor wire. An ascent through a cloud with supercooled water can cause ice to form on the sensor, effectively deadening the response. Balloons can be difficult to launch when sustained ground wind speeds are above 10 knots, although there are techniques that can push the limit to nearly 25 knots. If the ground station is located in a hilly area, high winds aloft can cause loss of signal as the balloon drops out of line of sight due to the large horizontal displacement.

The thermosonde is a well-known, reliable, and reasonably affordable instrument used to perform in situ measurement of optical turbulence. Over the years, it has compared well with other means of measurement, and it is unique in the precision to which it can measure and locate regions of high altitude optical turbulence. An important point to remember is that it is not providing a true vertical profile, but rather a profile along a path that can extend well away from the launch site.

Authored by
M. Chun
University of Hawaii, Honolulu, Hawaii

R. Avila
Universidad Nacional Autonoma de Mexico, Morelia

G. Jumper
Air Force Research Laboratory, Hanscom AFB, MA

A. Tokovinin
Cerro Tololo Interamerican Observatory, Casilla 603,
La Serena, Chile

T. Travoullion
Thirty Meter Telescope Project, California Institute of
Technology, Pasadena, California

R. Wilson
Physics Department, Durham University, Durham UK

A. Ziad
Laboratoire Universitaire d'Astrophysique, Universite de Nice

*Calibration of U.S. thermosondes is a combination of radiosonde calibrations by the manufacturer to instrument circuit calibrations as discussed in Murphy et al. (2007).

CHAPTER 5

Seeing by Site Monitors Versus Very Large Telescope Image Quality

Since the commissioning of the Very Large Telescope (VLT) it has been known that the image quality delivered by the telescopes is better, and often much better, than predicted by the seeing monitor. The advent of new sensitive instruments to measure the optical turbulence profile of the atmosphere over Paranal has finally allowed us to understand the origin of this discrepancy: the presence of a highly turbulent layer so close to the ground that it is seen by the seeing monitor, but not by the VLT unit telescopes. In this chapter we tell the story of this elusive surface layer.

5.1 The Inconvenient Discrepancy

It has been known since the commissioning of the VLT that the image quality delivered by the Unit Telescope (UT) is at times significantly better than the seeing measured by the Differential Image Motion Monitor (DIMM). The difference is not subtle. Already in 1999 the careful observations made with the test camera during the commissioning of UT2 revealed an alarming discrepancy between the UT2 image quality and the DIMM seeing, with an average DIMM–UT2 difference of ~0.2". During these tests that lasted several nights, UT2 was pointing at the same region of the sky and through the same filter as the DIMM, so there was no straightforward explanation for the lack of agreement. A dramatic manifestation of the discrepancy between the DIMM and the UT's is given by the time evolution of seeing on Paranal shown in Fig. 5.1. The DIMM seeing has degraded considerably over the past 17 years from a median value of 0.65" in 1990 to more than 1.1" in 2007. On the other hand, the right panel of Fig. 5.1 shows that the image quality delivered by FOcal Reducer and Low Dispersion Spectrograph 2 (FORS2) and Infrared Spectrometer and Array Camera (ISAAC) seem to have improved with time. This result could however be a selection effect since some of the PI's requested special seeing conditions. We note in passing that the La Silla seeing has also slightly degraded in a similar period (http://www.eso.org/astclim/ paranal/seeing/).

We have been puzzling for a long time about the origin of this rather inconvenient discrepancy, but it has not been until recently, with the deployment of new sensitive instruments in the context of the Extremely Large Telescope (ELT) site testing campaign, that we have finally been able to draw a coherent picture. This chapter tells the story of the seeing on Paranal.

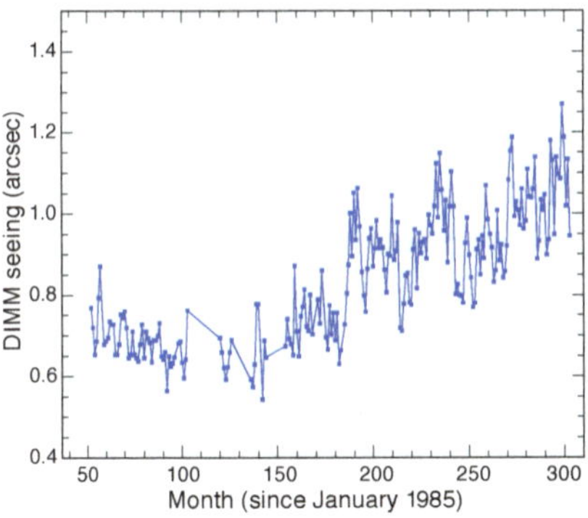

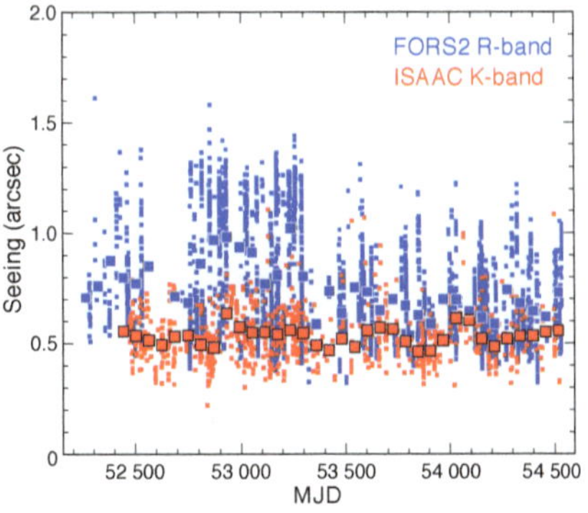

Figure 5.1 Left: Evolution of DIMM seeing on Paranal since 1989 (http://www.eso.org/gen-fac/pubs/astclim/ paranal/seeing/singstory.html). Right: Evolution of the FORS2 image quality in the *R*-band and of ISAAC image quality in the *K*-band since January 2002 (MJD = 52 275). The big squares show the averages over 2-month bins.

5.2 New Data

5.2.1 FORS2 Imaging Data

A wealth of data has accumulated since the commissioning of UT2. For example, the Quality Control process (QC) systematically logs delivered image quality from several instruments together with environmental parameters such as wind speed and direction, air temperature, telescope position, and DIMM seeing. The most complete dataset for image quality is the one for FORS2, which will be used here. Figure 5.2 shows the relation between DIMM seeing and UT2 image quality measured during regular FORS2 operations. The FORS2 data has been corrected for wavelength and airmass using the standard formulae based on an infinite outer scale assumption. Only images taken at airmasses less than 1.5 and exposure times between 30 s and 300 s were used. The FORS2 measurements are automatically obtained using many objects on each frame, but only values for which the image quality dispersion is less than 0.1" rms have been retained.

The correlation between DIMM and FORS2 reproduces the trend observed with the test camera during the commissioning of UT2. The mean DIMM seeing is 0.81" while the mean FORS2 image quality is 0.65", so on average the DIMM overpredicts image quality by about 0.16", similar to the value of ~0.2" measured with the test camera. It may be tempting to apply a rule-of-thumb correction of ~0.15" to go from DIMM seeing to UT image quality (at similar airmass and wavelength), but one should notice that for very good seeing conditions the DIMM seeing may be better than the FORS2 image quality, while under very bad conditions, the DIMM may indicate a seeing more than 1" worse than FORS2. So it is important to understand the origin of the discrepancy between DIMM seeing and UT image quality.

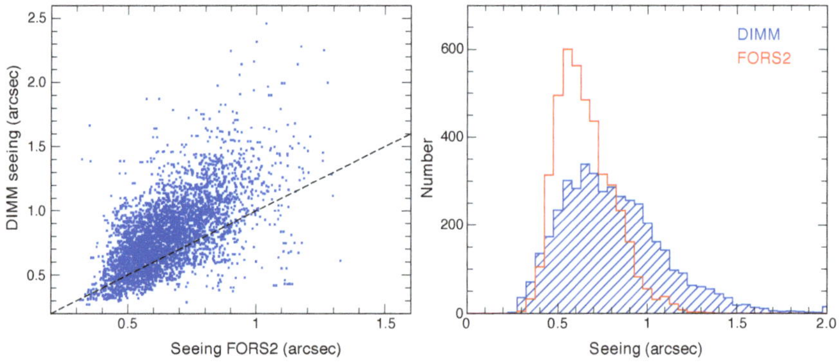

Figure 5.2 Relation between DIMM seeing and image quality measured by FORS2 between 2004 and 2006. In the left plot is shown a point plot, and the dashed line indicates DIMM = FORS2 seeing. In the right plot, the measurements are shown as histograms of the seeing values for both instruments.

5.2.2 Active Optics Shack-Hartmann Data

Real-time quality control of service-mode data at the observatory requires a reliable way of assessing whether the data complies with the seeing requirements set by the PI's. While this is straightforward for imaging data, it is not so for spectroscopy. Thus, the observatory operations staff typically rely on the FWHM of the stars in the guide probes to estimate the seeing, which seems to be sufficiently accurate for Observing Blocks (OB) validation purposes. The possibility of systematically using the sizes of the Shack-Hartmann (SH) spots of the active optics system of the UT's has the advantage that the sizes are routinely logged by the telescope control software and therefore could provide a readily available real-time estimate of the image quality. Figure 5.3 shows, in the left panel, a comparison between the image quality measured by FORS2 and the SH for about 750 simultaneous observations between 2002 and 2007 (blue dots). The Shack-Hartmann data were corrected by the diffraction on the 40 cm diameter SH apertures (0.35") (green dots). The right panel shows the corresponding histograms. The median image quality measured by FORS2 is 0.64", and 0.63" by the (corrected) SH spots; both histograms are seen to coincide very nicely. The figure shows that indeed the (diffraction-corrected) size of the SH spots provides an excellent proxy for image quality. Using the SH information we now have access to a much larger dataset to compare DIMM seeing with UT image quality.

5.2.3 Atmospheric Turbulence Profiles—C_n^2 (h)

Modern site characterization campaigns aim at determining the vertical turbulence profiles of the atmosphere at each site. The most direct way of doing

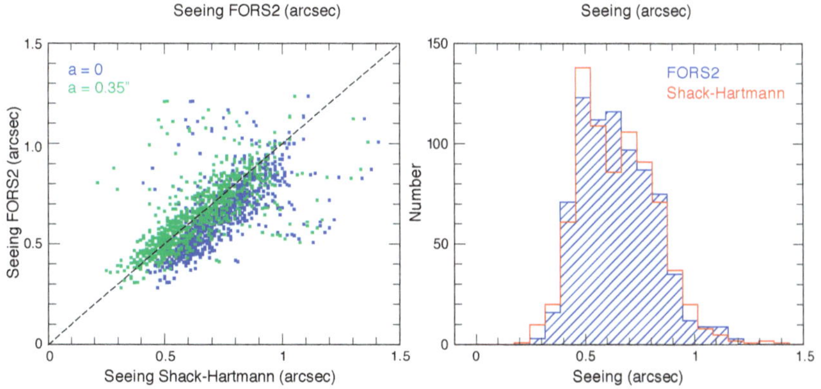

Figure 5.3 Left: Relation between the image quality delivered by UT1 estimated using the Shack-Hartmann (SH) spots of the active optics sensor, and the value determined on long (30 s–300 s) exposures with FORS2 in the R-band. The data have been normalized to airmass 1.0 and 500 nm wavelength. The blue points are the original SH data, and the green points show the values corrected for a lenslet aberration of 0.35". The dotted line corresponds to FORS2 = SH. Right: Measurements presented as histograms for FORS2 and the SH.

this is to fly balloons equipped with very sensitive sensors that can measure the temperature and wind speed fluctuations as a function of altitude. Of course these experiments are costly and cannot provide real-time diagnostics of the conditions on a given night. Thus, a number of techniques have been developed to do the job from the ground. The Multi-Aperture Scintillation Sensor (MASS) is a compact single-star instrument that measures scintillation on four concentric zones of the telescope pupil using photomultipliers (Kornilov et al., 2003; see also Chapters 3 and 4). A statistical analysis of these signals measures the vertical profile of turbulence C_n^2 (h) in six layers at 0.5, 1, 2, 4, 8, and 16 km above the telescope. A MASS unit developed at the Sternberg Institute (Moscow) under joint ESO-CTIO funding observed continuously on Paranal between 2004 and June 2007. In addition to its low-altitude resolution (about half the layer altitude, i.e. $\pm$ 250 m at 500 m and $\pm$ 8 km at 16 km), a distinct disadvantage of MASS is that it is blind to turbulence close to the ground (the ground layer), which produces little scintillation. We will show below, however, that for the purpose of understanding the discrepancy between DIMM and UT's, this turns out to be an important advantage.

5.2.4 The Ground Layer

A straightforward application of the MASS data is to integrate the profiles to measure the seeing. This is shown in the left panel of Fig. 5.4 where the seeing measured by the MASS is compared to the seeing observed simultaneously by the DIMM. As expected, the MASS systematically underestimates the 'real' seeing because it does not "see" the turbulence that is close to the ground. Therefore, Fig. 5.4 tells us that a turbulent layer located well below 500 m from the ground produces a significant fraction of the seeing over Paranal. This is something that was already known from previous experiments involving microthermal sensors on a mast (Martin et al., 2000). What is new is that we have a substantial body of simultaneous observations, which we can use to quantify the contribution of the ground layer with excellent time resolution.

The seeing ε is linearly proportional to wavelength and inversely proportional to the Fried parameter r_0. If we assume that the atmosphere has only two turbulent layers, a ground layer (GL) and a high-altitude layer (HA), then the total seeing is:

$$\varepsilon_{Tot}^{5/3} = \varepsilon_{GL}^{5/3} + \varepsilon_{HA}^{5/3} \tag{5.1}$$

Using this equation we can estimate the ground layer component since DIMM measures ε_{Tot} and MASS measures ε_{HA}. The result is presented in the right panel of Fig. 5.4 that shows the histogram of ground layer seeing. The mean ground layer seeing on Paranal is 0.72" with a rather large dispersion of 0.36" (σ) indicating that the ground layer varies significantly with time. So the comparison between MASS and DIMM tells us that a substantial fraction of the seeing on Paranal originates in turbulent layers below 500 m altitude. The resolution of MASS does now allow us to say more, but there are other instruments that can get us closer to the ground.

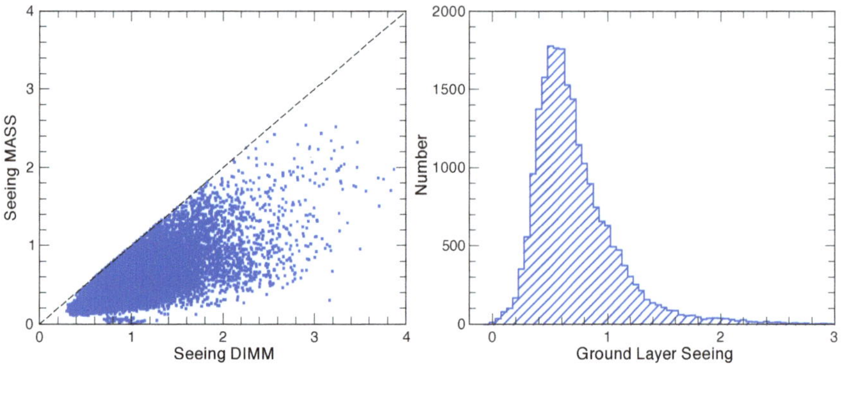

Figure 5.4 Left: Comparison between DIMM and MASS seeing (arcsec). The small fraction of events with MASS overshoots have been removed for computing purposes. The MASS seeing is significantly smaller indicating a significant contribution from a the ground layer (h < 500 m). Right: Histogram of the ground layer seeing contribution.

5.2.5 SLODAR

The Slope Detection and Ranging instrument (SLODAR) uses an optical triangulation method on double stars to measure the atmospheric turbulence profile (Butterley et al., 2006; see also Chapter 4). SLODAR, that has had several observing runs on Paranal since 2005, gives C_n^2(h) for eight layers with a resolution between 50 m and 100 m, depending on the separation of the double star and the zenith angle. While MASS measures the atmosphere between 0.5 and 16 km, SLODAR measures below 1 km, so both instruments are nicely complementary (although as stressed above MASS has a much lower vertical resolution). Figure 5.5 shows the distribution of the ratio of the contribution of the first (SLODAR) layer (h < 94 m) to the total ground layer turbulence determined by combining together DIMM, MASS, and SLODAR data taken simultaneously (Lombardi et al., 2008). This technique allows the restoration of unified profiles from the surface (or platform) of the observatory to the high atmosphere. The application of this technique to a large database from Paranal Observatory gives detailed statistics of the evolution of the turbulence and the seeing at Paranal. The plot shows that most of the time the ground layer turbulence is concentrated below 94 m. The median value of the distribution is 0.86" and the mean 0.82", but the distribution is heavily skewed toward large values indicating that conditions where the ground turbulence is above 94 m are quite rare. The strong turbulence at a mean altitude of ~50 m revealed by these observations suggests that the inconvenient discrepancy could be explained if much of this turbulence is in fact below ~20 m, so it is seen by the DIMM but not by the UT's. Some evidence in support of this hypothesis comes from the Lunar Scintillometer (LuSci) developed by Tokovinin (2007). LuSci allows the ground turbulence to be measured with a resolution of

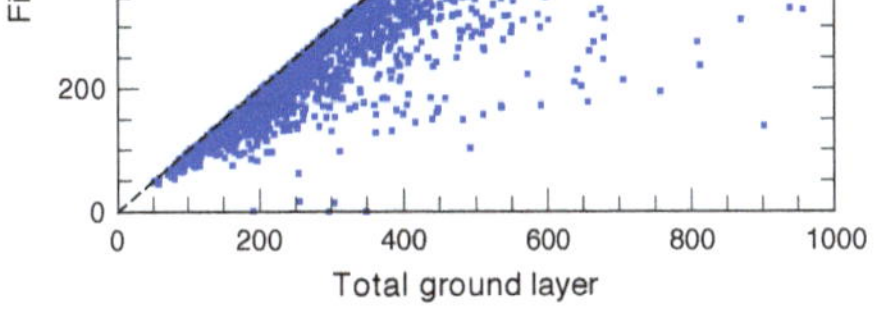

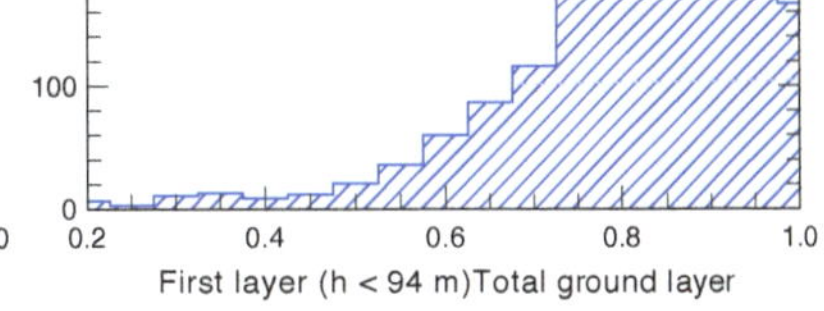

Figure 5.5   Left: C_n^2 (h) (h < 94 m) versus total ground layer turbulence measured by SLODAR. Right: Fraction of the total ground layer turbulence that comes from the first SLODAR layer. Most of the time the ground layer contribution is dominated by the h < 94 m component.

~10 m from observations of the lunar disc. A very preliminary LuSci test run at Paranal in December 2007 indicates that a substantial fraction of the ground layer turbulence is indeed lower than ~15 m above the platform on Paranal. Hereafter we will refer to this (low) layer as the 'surface layer'.

5.3 Understanding the Inconvenient Discrepancy

We can test our hypothesis about the nature of the inconvenient discrepancy by correcting the DIMM seeing for the ground component using the MASS data and comparing the results with the UT image quality as measured by the SH spots. For this comparison we need to know the fraction of the total ground layer seeing contributed by the surface layer (h < 20 m). The SLODAR data tells us the average value is ~0.8. The best fit shown in Fig. 5.6 is obtained for lenslet aberration a = 0.35" and a mean surface layer fraction of 0.8. For these values the least squares fit (solid line) agrees within 1% with the X = Y solution, but the histograms for the two datasets do not overlap exactly. The best match is obtained for a surface layer fraction of 0.7. This is not surprising since, as we have seen above, the value changes with time, so assuming a constant is just an approximation. The generally good agreement between surface-layer corrected DIMM seeing and UT image quality, however, provides convincing evidence that the surface layer is indeed the most likely explanation for the inconvenient discrepancy

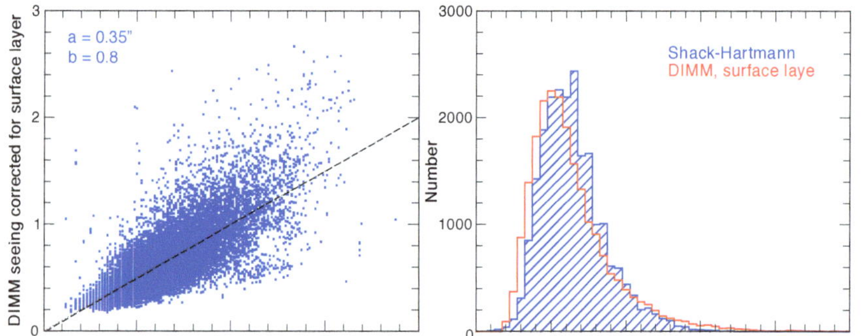

Figure 5.6 Left: Relation between DIMM seeing above the surface layer determined as described in the text, and the UT1 image quality estimated using the size of Shack-Hartmann (SH) spots of the active optics. The solid line shows a least squares fit to the data of slope 1.0. The best match of the two lines is obtained for an intrinsic SH spot size of a = 0.35", and a surface layer which contributes about 80% of the total ground layer seeing measured comparing DIMM and MASS. Right: Histograms of DIMM seeing corrected for surface layer and the SH image quality. While the mean values of the two histograms coincide, the overlap is best for 70% surface layer contribution.

5.3.1 When the Seeing is Bad

The automated Vaisala weather tower on Paranal provides continuous data that we can use to investigate the conditions that influence the presence and strength of the surface layer. Assuming that the DIMM/UT discrepancy, *ds* (see caption of Fig. 5.7), measures the strength of the surface layer, we can use the Vaisala data to check whether the surface layer strength correlates with Paranal environmental parameters. Figure 5.7 shows the wind-rose of Paranal color coded according to *ds* on the left panel, and by the difference in temperature between 30 m and 2 m on the right panel. The discrepancy is seen to be strongest when the wind comes from the NNE and from the SSE (with a broad distribution about these directions), while the temperature gradient is largest when the wind comes from the NE and SSE. This suggests that the bad seeing occurs when the wind blows warm turbulent air from nearby summits along the Atacama fault (which traces most of the road between the Panamerican highway and Paranal) over the top of the mountain. A temperature inversion of 0.5° C is present most of the time on Paranal and there is a weak trend of the DIMM/UT discrepancy increasing with the 2–30 m temperature difference, indicating that local conditions may play a role in determining the properties of the surface layer (e.g., by confining it to very low altitudes). An investigation of this aspect of the problem is underway, but is beyond the scope of this chapter.

5.4 Blown with the Wind

If our interpretation of the inconvenient discrepancy is correct, we expect the surface layer to have become increasingly important with time, but the

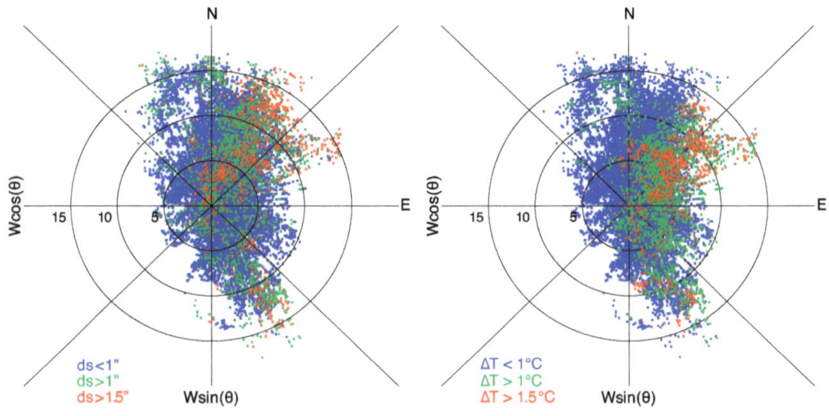

Figure 5.7 Left: The wind-rose of Paranal color coded by the discrepancy between DIMM seeing and UT1 image quality; ds = $(DIMM^{5/3} - SH^{5/3})^{3/5}$. Right: Wind-rose coded by temperature gradient between 2 m and 30 m above ground.

other components of the seeing to remain constant. Lombardi et al. (2008) examined this question using the combined DIMM+MASS+SLODAR data taken simultaneously between 2005 and 2007 and found that about 70% of the total ground layer turbulence is concentrated in the surface layer.

Although there is a worsening of the turbulence of the ground layer at Paranal, this worsening is almost concentrated in the first 50 m above the ground in a way that is not significantly affecting the image quality of the VLT Unit Telescopes. Their results, reproduced in the right panel of Fig. 5.8, show that this is indeed the situation. The degradation of DIMM seeing on Paranal is seen to be completely due to changes in the ground layer, while, if anything, the high-altitude layer seems to be getting better. A similar result spanning a longer time interval, albeit with lower altitude resolution, is obtained comparing the DIMM and MASS data only (right panel in Fig. 5.8). The surface layer (which as we saw is the main component of the total ground layer) has become significantly stronger over the past four years, and from the evolution of the DIMM seeing shown in Fig. 5.1 we infer that this has been going on for the past 12–15 years.

If the surface layer is blown with the wind over Paranal, we expect the wind distribution to have changed over the years. The evolution of the wind pattern on Paranal since 1985 is shown in Fig. 5.9. The analysis of the astroclimatology data (www.eso.org/astclim/paranal) shows that indeed the frequency of 'bad winds' (NE, NNE, SSE, and S) has increased over the past 15 years. However a link to regional changes of the wind pattern, possibly due to climate change, has not been demonstrated. Fortunately, at Paranal the turbulence blown by the wind is very close to the ground, so telescopes high above the ground don't see it. Unfortunately, telescopes close to the ground do!

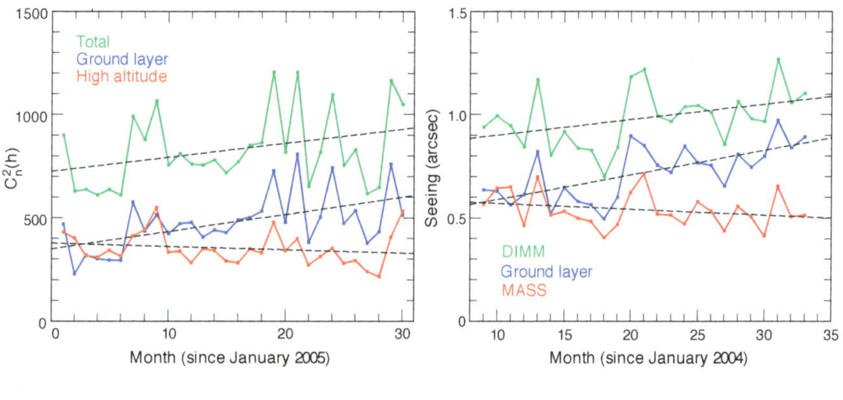

Figure 5.8 Left: Evolution of the components of atmospheric turbulence over Paranal (C_n^2) between 2005 and 2007 determined by combining DIMM, MASS, and SLODAR data (from Lombardi et al., 2008). Right: Evolution of the seeing components determined combining the DIMM and MASS data as described in the text.

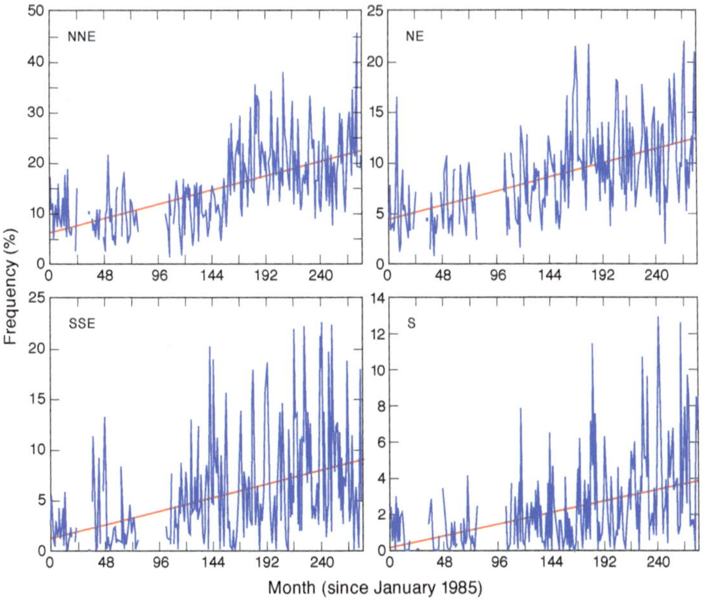

Figure 5.9 Evolution of the wind patterns over Paranal since 1985. The frequency of NE and NNE winds has increased dramatically since just about the time VLT commissioning started. The S and SSE wind fluctuations have increased such that during some months the frequency of these winds is also dramatically increased.

Figure 5.10 View of the Galactic Centre above La Silla with the domes of the 3.6-m telescope and the CAT illuminated by the setting Moon.

5.6 Conclusions

We can safely draw two quite strong conclusions about the evolution of seeing on Paranal:

- The discrepancy between the seeing measured by the DIMM and the image quality delivered by the VLT Unit Telescopes, and the notable degradation of DIMM seeing observed over the past 15 years have a common origin: the presence of a thin, time variable turbulent layer—the surface layer—over the mountain that is seen by the DIMM, but not by the UT's.
- The surface layer is strongest when the wind blows from the NNE and from the SSE. These winds have become increasingly frequent over the past 15 years explaining why the surface layer appears more and more often.

Site testing campaigns must pay close attention to the surface layer through the use of micro-thermal towers, or sensitive astronomical instruments such as SLODAR and LuSci. Extensive campaigns on existing observatories now underway should be intensified and the results cross-correlated, especially if different techniques are used. Close attention must be paid to the local orography, and the effects of changes in the prevailing winds modelled. Seen through the light of modern site testing techniques and global climate change, it is sobering to realize that our ancestor's conventional wisdom—put thy telescope as high above the ground as possible—is still right!

Authored by
M. Sarazin and J. Melnick
European Southern Observatory, Garching

J. Navarrete and G. Lombardi
European Southern Observatory, Chile

Part III
Adaptive Optics

CHAPTER 6

Introduction to Adaptive Optics:
The Quest for Image Quality

This short review is not intended as an introduction to Adaptive Optics (AO). The reader is referred to textbooks (Hardy, 1998; Roddier, 1999; Tyson, 1998, 2000) and to an on-line material. Here we only introduce basic principles and terminology to help understanding the following text. The subject of this chapter is well covered in the literature.

The purpose of AO is to compensate atmospheric distortions of the wave-front in real time, thus improving the image quality and resolution of ground-based telescopes. AO has many applications outside astronomy, which are not considered here (e.g., improving energy concentration in military systems or removing eye aberrations in ophthalmology). Despite similarity of terms, adaptive optics is not to be confused with active optics, where telescope aberrations are measured and corrected in real time, but too slowly to compensate the turbulence. These two techniques are indeed similar; their difference is just in the speed of correction.

Atmospheric distortion is measured by the Wave-front sensor (WFS) using the light of a guide star (GS) as a probe (Fig. 6.1). The GS can be natural (NGS)—a real star or small planet—or artificial. An artificial GS is normally created by scattering laser light in the air, so-called Laser Guide Star (LGS). As to the WFSs,

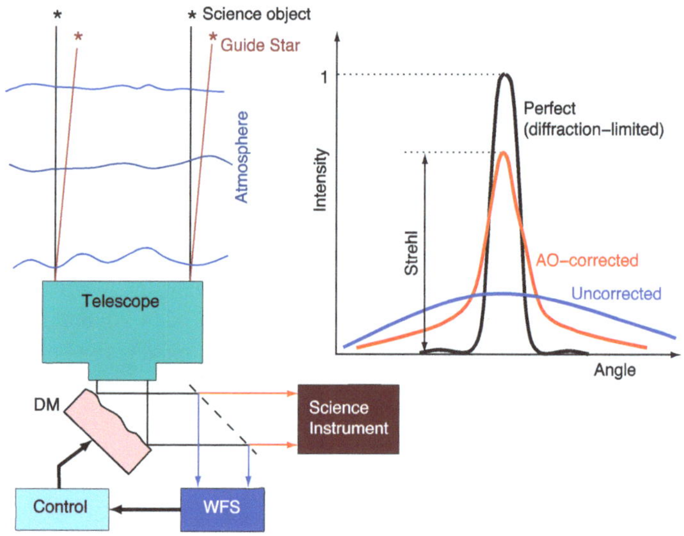

Figure 6.1 Principle of AO operation and schematic representation of the PSFs.

they use different principles for wave-front measurements, which will not be detailed here. Of importance are global WFS parameters: spatial and temporal resolution and its efficiency in using photons from the GS to measure wave-front distortions with the lowest possible noise.

The corrective element of an AO system is usually called deformable mirror (DM). A thin mirror driven by piezo-electric actuators can create phase distortions, which compensate turbulent phase, so that upon reflection from such a DM the wave-front will be more flat than before entering the AO system. Corrective AO elements use varying technology; they can even be transmissive rather than reflective. The physical size of the DM can be quite small compared to the corrected wave-front, so the latter is projected on the DM by some re-imaging system, which includes, of course, the telescope. For the purpose of this review, only such basic DM parameters as the total number of actuators N_{act}, maximum achievable amplitude of correction, stroke, and maximum speed of correction matter. The distance between adjacent actuators, as projected back on the wave-front, is called actuator pitch d_{act}. Obviously, for a telescope diameter D, $N_{act} \approx (D/d_{act})^2$.

The spatial resolution and number of elements in the WFS and DM are usually matched (but not necessarily equal). It makes no sense to have a DM with high actuator count if we do not get enough information from the WFS to control them. The control loop that processes the WFS signals before sending them to the DM implements rather complex algorithms. These algorithms, together with the hardware parameters, determine the final (or effective) number of controlled degrees-of-freedom (e.g., actuators) and the speed of the response. The time interval between successive WFS measurements is called loop cycle t_{loop}, typically ranging from 1 ms to 4 ms. The reaction of the DM to these data (hence the actual speed of turbulence correction) can take up to 10 t_{loop}, depending on the control algorithm. The correction speed is limited by various delays in the system, which conspire to make it unstable, if controlled faster.

When the WFS measures the incoming wave-front distortions directly and the corrections are sent to the DM, the AO system works in the open loop. Such a system requires a good accuracy of its main components, WFS and DM. All current AO systems work instead in closed loop, when the WFS measures the wave-front already corrected by the DM. This has a huge advantage that small errors in either WFS (e.g., non-linearity) or DM (e.g., hysteresis) get corrected at next loop iterations. The WFS eventually measures only small residual distortions, i.e., works as a null sensor. The AO system in Fig. 6.1 is of the closed-loop variety. Otherwise, the beam-splitter which sends part of the light to the WFS would be located before the DM. Quite often light is divided between WFS and science beams by wavelength, e.g., visible light goes to the WFS, the infra-red (IR) light to the science channel. Atmospheric distortions are nearly achromatic; therefore wave-front correction in the visible light is also good for the IR.

The final correction quality is evaluated by the residual wave-front variance σ^2. When this variance is small compared to the imaging wavelength λ, the resulting Point Spread Function (PSF) is nearly diffraction-limited. The ratio of the maximum PSF intensity to that of an ideal PSF is called Strehl Ratio (S). It can be shown that

$$S \approx \exp[-(2\pi\sigma/\lambda)^2], \qquad (6.1)$$

which means that S and σ^2 are closely related. In real AO instruments the correction is never perfect, $S < 1$. Roughly, S is the fraction of the light contained in the limited PSF core of the width λ/D or slightly larger. The remaining $1 - S$ fraction of the light forms the seeing-limited PSF halo of half-width λ/r_0 or somewhat less (Fig. 6.1). Such core-halo structure of the corrected PSF is typical for nearly all AO systems.

The terrestrial atmosphere is thick (scale-height ~8 km), so even for a large 10-m telescope the atmospheric light path resembles a very thin cylinder (aspect ratio 1:800). Looking at different directions, the path will be different. The distortions measured by the WFS and "seen" by a star some distance away are not the same, therefore the quality of AO correction degrades with the distance from the GS (anisoplanatism).

Optical distortions in the whole atmospheric volume (in three dimensions) can be measured, at least in principle, using several crossed beams. This approach, called turbulence tomography, requires several GSs and WFSs and works in analogy with medical tomography where a patient's body is sounded with several beams coming from different directions. Correction in 3D can also be made with several DMs conjugated optically to different altitudes. Such Multi-Conjugate AO (MCAO), envisioned long time ago by R.H. Dicke and J. Beckers, now becomes a reality.

In modern astronomy, different flavors of AO are now popular, each with its own acronym. Classical AO with just one NGS and one DM conjugated to the ground level is often called Single-Conjugate AO (SCAO), to distinguish it from MCAO. The Multi-Object AO (MOAO) is a method similar to MCAO, except that the information obtained from tomography is now used to correct each object individually with its own dedicated DM. MOAO can only work in open loop, because the light reflected by the DMs is not sensed. Yet another use of tomography would be to correct only the strongest low-altitude turbulence with one DM conjugated near the ground. Such correction, although partial, will be valid over a wide field, giving this Ground-Layer AO (GLAO) technique certain attractiveness. Finally, the acronym ExAO stands for Extreme AO, meaning a very high degree of correction ($S \approx 1$). ExAO is planned mostly for imaging extra-solar planets.

We begin by considering atmospheric effects for the simplest case of SCAO in Section 6.1. In the next Section 6.2 the 3-dimensional character of atmospheric distortions relevant to LGS and MCAO is considered. Finally, practical aspects of using data on atmospheric turbulence for AO are treated in Section 6.3.

6.1 Performance of Classical AO

6.1.1 Atmospheric Parameters Relevant for AO

Wave-fronts are distorted by turbulence at all spatial scales from millimeters to kilometers. There is no preferred scale of "turbulence cells", although this term can still be found in the literature. The strength of distortion depends on the scale in a smooth way which, for many purposes, is adequately described by the Kolmogorov model. This model predicts that the variance of the difference in phase between two points on a wave-front separated by a transverse vector $\vec{r}$ is

$$D_\phi(\vec{r}) = \langle[\phi(\vec{x}+\vec{r}) - \phi(\vec{x})]^2\rangle$$
$$= 6.88\,(|\vec{r}|/r_0)^{5/3}. \tag{6.2}$$

The quantity $D_\phi(\vec{r})$ is called phase structure function and is measured in square radians. The coefficient r_0 in equation 2 is called the Fried parameter and is measured in meters. At a distance of r_0, the rms phase difference reaches 2.6 rad, or $\lambda/2.4$. The variance equals 1 rad^2 at $r = (6.88)^{-0.6}\,r_0 = 0.314\,r_0$.

Atmospheric distortions of wave-front are nearly achromatic (neglecting small dispersion of the air), hence $D_\phi \propto \lambda^{-2}$ and $r_0 \propto \lambda^{1.2}$. For imaging, r_0 sets the characteristic scale of phase distortions because smaller wave-front details produce only effects $\ll\lambda$. An AO system must have an actuator pitch $d_{act} \sim r_0$ to work well. The dependence of r_0 on λ means that AO correction is much easier in the IR than in the optical because the actuator count $N_{act} \sim (D/r_0)^2 \propto \lambda^{-2.4}$.

Temporal evolution of distortions defines how fast an AO system must be to correct them. If we follow phase distortions at one point and measure the time after which the difference reaches 1 rad r.m.s., this time τ_0, AO time constant, will be

$$\tau_0 = 0.314\,\frac{r_0}{\bar{V}}, \tag{6.3}$$

where $\bar{V}$ is the wind velocity averaged over the altitude. The formula is easy to understand if the wave-front moves as a whole with translation speed $\bar{V}$. This assumption (Taylor hypothesis) is valid for individual parts of the atmosphere, but is not valid for the wavefront at the ground, produced by the cumulative effect of many turbulent layers at different altitudes h. Taking advantage of the power-law (2), it can be shown that the average wind speed

$$\bar{V}^{5/3} = [\int C_n^2(h)|\vec{V}(h)|^{5/3}\mathrm{d}h] / [\int C_n^2(h)\mathrm{d}h] \tag{6.4}$$

can be used in (3) for evaluating errors associated with the speed of the AO system. Here $C_n^2(h)$ is the vertical profile of the refractive index structure constant, called here for brevity turbulence profile (TP), and $\vec{V}(h)$ is the profile of the wind speed. The integration in (3) is carried from the telescope to the upper limit of the atmosphere.

The phase difference between two viewing directions separated by the angle θ (at one point on the wave-front) is caused by the spatial shifts θh between the sources, increasing with altitude h. This difference reaches 1 rad r.m.s. at $\theta = \theta_0$, where the isoplanatic angle is defined as

$$\theta_0 = 0.314 \frac{r_0}{\bar{h}} \tag{6.5}$$

and $\bar{h}$ is the effective altitude of turbulence calculated in a similar way as $\bar{V}$.

The atmospheric parameters relevant to AO, r_0, τ_0 and θ_0, are defined above for viewing at zenith (see Preface). For some zenith distance γ, $r_0 \propto \sec \gamma$ and $\theta_0 \propto \cos \gamma$. The variation of τ_0 with zenith distance is not so simple, it can either increase or decrease with γ, depending on the wind direction.

6.1.2 Quality of AO Correction

The average phase variance σ^2 across circular telescope aperture of diameter D equals

$$\sigma^2 = c \, (D/r_0)^{5/3}, \tag{6.6}$$

where the coefficient $c = 1.03$ without AO correction. When only overall tip and tilt are corrected, $c = 0.134$. In other words, 87% of the variance comes from tip-tilt (tt).

By increasing the correction order further, we keep reducing the coefficient c. There are approximate formulas to calculate residual error as a function of the number N of corrected parameters. The exact coefficient depends, however, on the nature of those parameters. For a simple control of N actuators, $c \approx 0.335N^{0.833}$, while if N Zernike modes are corrected, $c \approx 0.294N^{0.866}$ (valid for $N > 10$). The best effect (smallest c) is achieved by correcting so-called Karunen-Loeve modes.

The residual variance calculated by (6) with appropriate coefficient c is called fitting error and describes the amplitude of small-scale distortions left in the wave-front after AO correction. It can be re-phrased as $\sigma_{fit}^2 = (N_0/N)^{5/6}$, where N is the actual number of actuators and $N_0 = 0.27(D/r_0)^2$ is the actuator count required to reach acceptable correction $\sigma_{fit}^2 = 1$ rad. The corresponding actuator pitch will be $d_{act} \approx 0.27^{-0.5}r_0 \approx 1.9r_0$. The Strehl ratio will be only $S = e^{-1} = 0.368$. This regime of partial correction corresponds to the maximum gain in resolution for a given number of controlled parameters N. By choosing a smaller pitch $d_{act} = r_0$, we will obtain $\sigma_{fit}^2 = 0.34$ rad^2 and $S = 0.71$.

The fitting error is not the only source of residual AO errors, hence the above estimate of the Strehl ratio is optimistic. Other atmospheric errors are caused by the time constant τ of the control loop (remember that $\tau \sim 10\, t_{loop}$) and anisoplanatism at angular distance θ from the GS. The resulting error will be

$$\sigma_{AO}^2 \approx \sigma_{fit}^2 + \sigma_{time}^2 + \sigma_{aniso}^2 =$$

$$0.335 N^{-0.833} (D/r_0)^{5/3} + (\tau/\tau_0)^{5/3} + (\theta/\theta_0)^{5/3}. \qquad (6.7)$$

In fact (7) slightly overestimates the residual error, hence the approximate sign.

So far, we have not considered the measurement noise in the WFS. The number of photons received from the GS during one loop cycle is proportional to $N_{ph} \propto d_{act}^2\, t_{loop}$ (we assume that WFS elements have the same size as the actuator pitch). Translating this into the dependence on the atmospheric parameters and wavelength, we get $N_{ph} \propto [r_0(\lambda_0)]^3 (\lambda/\lambda_0)^{3.6} \overline{V}$, where $r_0(\lambda_0)$ defines the seeing. Note that the sensitivity of AO is a very strong function of seeing, but it does not depend at all on the telescope diameter D. The very purpose of AO was to free astronomers from the seeing but, paradoxically, AO makes this dependence even stronger.

If we want operate an AO system with NGS, a sufficiently bright star has to be found at a distance no more than θ_0 from the science object. The fraction of sky where such stars are available is called sky coverage (SC). Calculations show that the SC can reach few percent in the K-band ($\lambda = 2.2\ \mu m$), but becomes hopelessly low at shorter wavelengths. This is why LGSs are needed in the astronomical AO.

6.2 AO with Laser Guide Stars

6.2.1 Cone Effect

First LGSs were created by simple scattering of laser light in the air and are called Rayleigh LGS because the Rayleigh scattering (together with aerosols) is the main mechanism which sends laser photons back to the telescope. Only a tiny fraction of emitted photons is received back, making this and all other types of LGS very inefficient. High laser power is needed to get enough light for AO correction.

The second type of LGS is called sodium because the scattering mechanism is resonant excitation of sodium atoms in the mesosphere by laser light with $\lambda = 589$ nm (D2 line). The sodium layer at altitude ~ 90 km is created by ablation of meteorites. Its altitude, profile, and intensity change on time scales from minutes to years.

As LGS is created at finite distance from the telescope, its beam is not parallel, hence the atmospheric paths of LGS light and science object (star) are different (Fig. 6.2). If $\varphi(\vec{r}, h)$ is the phase distortion produced in the atmospheric layer at altitude h, the conic beam coming from the LGS altitude H projects this distortion

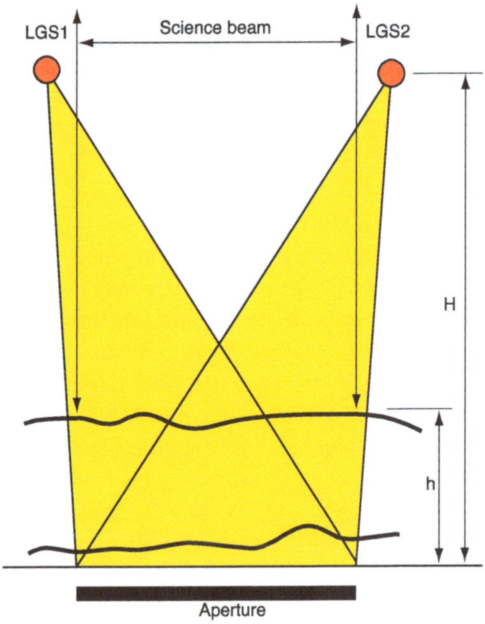

Figure 6.2 Cone effect and laser tomography (see text).

on the telescope pupil with a magnification (stretching) by a factor $H/(H - h)$ and a spatial shift θh, if the LGS is located at an angular distance θ from the science object:

$$\varphi_{\mathrm{LGS}}(\vec{r}) = \varphi[\vec{r}(1 - h/H) + \theta h] \qquad (6.8)$$

The distortion $\varphi(\vec{r}, h)$ is seen by the science object in full, while the WFS sees its stretched and shifted version $\varphi_{\mathrm{LGS}}(\vec{r})$. This phenomenon is called cone effect or focus anisoplanatism. Part of the wave-front is not sensed by the LGS.

An AO system with LGS will work worse than in the case of the NGS. The wave-front error due to cone effect is described by the parameter d_0 (in meters),

$$\sigma^2_{\mathrm{cone}} = (D/d_0)^{5/3}. \qquad (6.9)$$

If atmospheric turbulence were concentrated in a thin layer near the ground, there would be no cone effect ($d_0 = \infty$) and no anisoplanatism ($\theta_0 = \infty$). Both parameters are related to the effective thickness of turbulence and, naturally, are correlated. As a first approximation, one may use $d_0 \approx 2.91\theta_0 H$, where H is the altitude of the LGS.

Astronomers prefer sodium LGS over Rayleigh LGS simply because $H = 90$ km gives a much smaller cone effect than $H \sim 15$ km typical for Rayleigh LGS. However, even with sodium LGS the cone effect becomes prohibitively large at large telescopes or at short wavelengths. AO systems with single sodium LGS work on 10-m telescopes in the IR, but will fail at a 30-m telescope.

6.2.2 Tilt Anisoplanatism

Ironically, LGS AO systems still need natural stars to work properly. The position of the LGS on the sky fluctuates (with respect to stars) because the turbulence introduces random motion of the projecting (up-link) laser beam. Therefore, at least one NGS is needed to measure the tip and tilt. Such tt NGS, however, can be fainter (because it uses full telescope aperture rather than sub-aperture of size d_{act}), because atmospheric tilts are correlated over larger angular distances than high-order aberrations (larger than θ_0), and because the tilt is slower. As a result, the sky coverage (SC) of LGS AO is much better compared to NGS AO, but it is still very low in the visible.

The SC of LGS AO does not improve with telescope diameter. Although a larger D collects more photons from the NGS, the requirement to measure tilt with errors smaller than λ/D annulates this gain. However, atmospheric tilts at large-aperture telescopes are drastically reduced by the finite turbulence outer scale. Even uncompensated tilt makes LGS AO resolution much better compared to seeing. Such observational mode has been tried at the Very Large Telescope (VLT).

There are various ways to alleviate the problem associated with tt NGS. For example, to measure tt in the IR where there are more stars on the sky and more photons. Or to use several NGSs and combine this information cleverly, estimating the tilt in the direction of the science object. The most drastic solution would be to place a special dedicated LGS on top of the NGS just for correcting the turbulence in this direction. The NGS image will then become diffraction-limited, requiring much less photons to measure the tilt accurately. So far, this expensive solution has not been implemented.

6.2.3 Laser Tomography

The term laser tomography AO (LTAO) refers to situations where several LGSs are created and the corresponding wavefronts are measured simultaneously with several WFSs. In analogy to the medical tomography, this permits to separate phase distortions coming from different altitudes or, in other words, to measure the atmospheric turbulent volume in 3 dimensions. This opens the way to correcting the intrinsic problem of LGS—the cone effect—because the stretching of wavefronts can be accounted for. Having this 3D information, we can calculate the phase correction required for the science object (project the 3D volume along the line of sight). This calculation can be done for several objects in the field, but in this case the corrections will be different, so we must use as many DMs as there are objects and operate in the open loop (MOAO). Better still, we can conjugate several deformable mirrors to different altitudes and correct the whole turbulent volume (MCAO).

There are obvious limitations to LTAO. The number of unknowns (phase perturbations) must be less than the number of measurements, which means that the 3D distortion is represented by few discrete layers and the number of these layers is no more than the number of LGSs, K. The tomographic resolution in altitude is thus proportional to K. The transverse resolution (finest detail on each layer) is, at best, on the order of Θh, where Θ is the angular radius of the LGS constellation and h is the vertical distance between the layers. Compared to the case of single LGS, LTAO improves the quality of correction, but this improvement is not magic and comes at a cost of increased system complexity. However, LTAO is mandatory for using LGS at large telescopes or at short wavelengths, because otherwise the cone effect would be too strong.

Real turbulence is distributed at all altitudes, instead of being concentrated in few layers. The minimum phase error of wave-front reconstructed with LTAO can be estimated as

$$\sigma^2_{\text{LTAO}} \sim (\Theta/\gamma_K)^{5/3}, \quad \gamma_K = r_0/\delta_K, \tag{6.10}$$

where Θ is the radius of the tomographic field and δ_K is the effective thickness of the atmosphere which can be calculated for a given turbulence profile and a given constellation of K LGSs. The angle γ_K can be called tomographic patch size (Tokovinin and Viard, 2001).

For realistic TPs, $\delta_5 \sim 0.5$ km. Comparing this to the effective thickness of the atmosphere for classical AO, $\bar{h} = 0.314 r_0/\theta_0 \sim 5$ km (assume $r_0 = 0.15$ m and $\theta_0 = 2''$), we see that LTAO with 5 LGSs opens up the field by a factor of 10.

In the case of MCAO, the residual errors also depend on the number and conjugation altitudes of the DMs. Even if we knew the 3D distortions perfectly, the residual error would be on the order of $(\Theta/\theta_M)^{5/3}$, where θ_M is the generalization of the isoplanatic patch size to the case of M correctors and Θ is the size of the corrected field. In reality, an MCAO system gets information from LTAO, so the residual error will be a combination of tomographic and correction errors, and will be larger than the LTAO error alone.

One caveat of LTAO is related to the tilt indeterminacy inherent to LGS. As the image of the LGS constellation can change its scale or stretch in one direction due to tilts, corresponding distortions of the science field cannot be determined from the LGS signals alone. Instead, at least 3 NGSs distributed in the field are now needed to constrain the field distortions which are related to the differential focus and astigmatism modes between the reconstructed layer (or between the DMs in MCAO). One may think that the sky coverage of LTAO becomes at affected, but in fact it is better than for a classical AO because the size of the field where three (instead of one) NGSs are to be found is now much wider.

6.2.4 Ground-layer AO

The strongest turbulence is usually concentrated near the ground (or inside the dome). By correcting selectively only this ground layer (GL), we will not reach the diffraction limit (uncorrected high-altitude turbulence is too strong), but will improve somewhat the resolution compared to natural seeing. However, this correction will be valid in a relatively wide field, making this ground-layer AO (GLAO) a valuable option for astronomy.

The best way to measure selectively GL turbulence is to use several guide stars in a ring configuration around the field and to average their signals. The ring radius Θ, actuator size d_{act} and the thickness of the corrected layer h_{GL} are related by the obvious geometry, $d_{act} \sim h_{GL}\Theta$. For a smaller field, we can correct a thicker layer and get a better resolution, so there is a trade-off here. In fact the correction does not drop abruptly above h_{GL}. Turbulence in the "grey zone" at $h \sim h_{GL}$ is partially corrected and is mostly responsible for the variation of the PSF across the field.

GLAO systems can use sodium LGSs deployed for LTAO and simply reconfigured for a wider field. Low-altitude Rayleigh LGSs would work even better. Even a single Rayleigh LGS at low altitude will be useful, because due to the cone effect it will sense preferentially the GL turbulence. The resulting performance is not optimum, but the resulting GLAO system will be rather simple. It was also suggested to rotate single LGS and WFS synchronously, fast enough to average the turbulence during one loop cycle, although this idea has not yet been put to test.

6.3 Use of Site Data for AO

Performance of AO systems depends critically on various parameters related to the atmospheric turbulence—seeing r_0, AO time constant τ_0 and isoplanatic angle θ_0. For LTAO, MCAO or GLAO, we also need to know the turbulence profile (TP) $C_n^2(h)$. AO requirements drive the development of the instruments to measure all these parameters.

Atmospheric parameters are used in AO in different ways.

- System design. The parameters and the range of their variation are used to dimension the AO system (number of actuators, guide stars, DMs), to calculate its expected performance and to address various trade-offs.
- Site selection. Considering that AO will be the standard mode on new extremely large telescopes (ELTs), its estimated performance drives the choice of the best sites. Seeing was the major site selection criterion for the classical telescopes, but now the emphasis shifts to the high-altitude turbulence and related parameters θ_0, τ_0. A site where these parameters are better will gain in the AO performance, even if it does not have the best seeing.

- Data reduction in AO is sometimes assisted by the knowledge of atmospheric parameters at the moment of observation. For example, measurement of the isoplanatic angle is helpful for predicting the PSF variation across the field (Britton, 2006). Parameters such as r_0 and τ_0 can be estimated in real time from the AO loop data and recorded together with the images. These estimates are more useful than the data from site monitors because turbulence changes in space and time. The measurement derived from the AO data refers to the time and viewing direction of the science object and is therefore most relevant.
- AO operation. Considering that the AO performance depends strongly on various atmospheric parameters, the moments of critical observations should be selected on the basis of those parameters, not just on seeing. Such advanced queue scheduling is not yet implemented, but it can substantially increase the science yield of large telescopes with AO. As different kinds of AO require different optimum combinations of parameters, the observing time can be allocated with increased efficiency.

To support AO operation, a site monitor should measure the three major AO parameters r_0, τ_0, and θ_0 reliably and in real time. Considering the statistical nature of those parameters, it does not make sense to require high measurement accuracy. In addition, the TP should be monitored. For GLAO, we need to know the TP with high resolution (20 to 100 m) only near the ground, but for predicting the LTAO performance the vertical resolution must be better than δ_K (~0.5 km) over the whole atmosphere. Considering again turbulence instability in space and time, the most useful TPs will be derived from the data of the LTAO system itself, rather than from the site monitors. A requirement to measure atmospheric parameters from the loop data should be set explicitly in the development of new AO systems.

Authored by
A. Tokovinin
Cerro Tololo Inter-American Observatory, La Serena, Chile

S. Businger
University of Hawaii, Honolulu, HI

Part IV
Modeling Optical Turbulence

CHAPTER 7

The "Missing Link" Between Meteorology and Astronomy

Unlike most other scientists, astronomers are uniquely dependent on the whims of nature to succeed. For centuries astronomers have studied the heavens with only elementary weather forecasting skills, perhaps in the form of reading the shape of approaching clouds or the hand of a barometer. In recent decades, through the use of satellite imaging and computer generated forecasts, astronomers have been able to gain some advance warning of inclement weather, which may dictate their ability to conduct nighttime observations. Nonetheless, even toward the end of the 20th century, this level of forecasting was rudimentary compared to what is possible today. We explain the unique interconnection between meteorology and astronomy, including the strategic importance of weather forecasting in the operations of modern observatories; we touch on the history of the Mauna Kea Weather Center and its unique niche in astronomy, the interdisciplinary synergies it relies upon, and the challenge of providing this service in the global astronomy arena.

7.1 Weather's Impact on Astronomy

As a scientific field, astronomy is unique in many ways. It is among the oldest intellectual endeavors in civilization, with roots that touch upon everything from religion (as the gods portended wars through eclipses), to geography (using the stars to prove the earth was round), to maritime applications (measuring latitude reliably anywhere at sea), to a deeper understanding of the origin and nature of the vast realm we live in—the universe. Perhaps the rich history behind astronomy is why it is so broadly endeared by society—we all can make a connection to astronomy at some level. Astronomy is unique in one other important way—for centuries astronomers have conducted their observations of the heavens through a fickle atmosphere that does not consistently pass radiation to the ground where they await with just their eyes or today, with sophisticated electro-optic systems. Indeed, compared to other scientists the modern astronomer is as susceptible to success or failure in conducting their observations as a corn farmer in Kansas is to growing a crop. For the chemist working in his/her lab to develop new compounds or the biologist growing new cultures (Fig. 7.1) to test advanced drugs, the weather has little career impact beyond the level of convenience they have driving to work each morning. But for the astronomer, even a patch of clouds in the "wrong" part of the sky can obliterate long-held plans to conduct research on some particular object in the sky. For the astronomy graduate student, this has even deeper implications. If a single target is the topic of one's thesis, and observing time is scheduled only semiannually, losing an observing run to poor weather can immediately inject a year of delay in the start of a young scientist's graduation and therefore career. Weather matters more to astronomy than perhaps any other field of science except, of course, meteorology.

Astronomers carefully choose the locations of new observatories, factoring in a range of parameters including cost, seeing, infrastructure, and—of course— the fraction of time that skies are covered by clouds. With major observatories having lifetimes ranging from 50-100 years, picking the right spot has long-term implications for the scientific success of an observatory, which involves massive permanent structures not conducive for transport to better sites. To demonstrate the magnitude of the effect, Fig. 7.2 shows the percent of time that clouds significantly interfere with observations at Mauna Kea—a site globally recognized as one of the premier sites in the world to conduct astronomy. Nearly 1/3 of the time astronomers spend on Mauna Kea is spent waiting for the skies to clear so they can conduct their research. While astronomers generally take this impediment to their field in a matter-of-fact manner (how can I change the weather?), it is tantamount to that chemist or biologist not being able to reach their research labs nearly 2 days each work week—a loss of data that would surely impact their research. Granted, these weather-related downtimes for astronomers are often seasonal (winter in Hawaii is generally the worst time for viewing), hence some long-term prediction of the probability of being cloudy is possible, but that is of little consolation for astronomers who are studying targets that exclusively rise in the winter skies.

For centuries astronomers have dealt with inclement weather by using

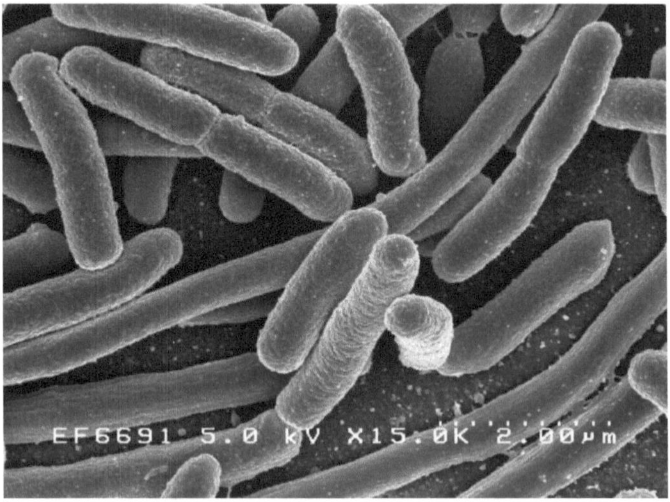

Figure 7.1 Microbes are visible under the microscope for a biologist to study regardless of the weather outside. Unlike almost all other scientists, astronomers are subject to the whims of the skies to perform their research.

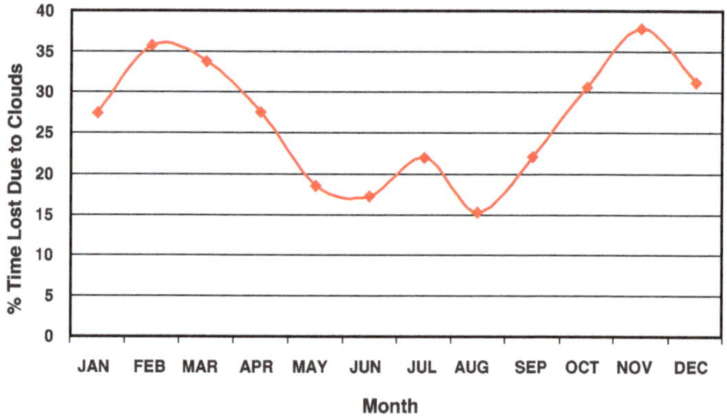

Figure 7.2 Clouds have a large impact on ground-based astronomical research, even at the best sites in the world, such as Mauna Kea, Hawaii, where roughly 1/3 of the time astronomical research is impossible due to limited atmospheric transparency.

essentially reactionary strategies as they adjust their observing plans after the "damage is done" to make up for lost research opportunities. Nonetheless there have been attempts to reduce the impact of poor weather on astronomy through various forecast methods. Figure 7.2 suggests that simplistic, climatologically driven seasonal forecasts that define long-term trends can be used. These of course have very limited value other than trying to select targets that are not up when the seasonal weather has been found to be the worst in the past. In recent centuries astronomers probably developed a keener sense of weather patterns than laymen and perhaps were among the most advanced weather forecasters as they learned to use different techniques to forecast the weather on a short-term basis. Tools we regard today as simplistic, like the barometer, changes in wind direction, or reading the form and motion of clouds, were key allies in the nightly battle with the weather that astronomers waged. A lowering atmospheric pressure was an indication of an approaching storm system. If confirmed by the sight of high cirrus and a switch in wind direction, this all suggested an approaching cold front, perhaps driving high-priority targets to be viewed first, before thick clouds spoiled observations for the next 2-3 nights.

Within the past 50 years, weather forecasting has improved remarkably thanks to advanced tools including satellites, vastly greater quantities of ground data, and perhaps most importantly, the advent of powerful computers capable of generating complex multidimensional models of the atmosphere days into the future. This has had a profound effect on society as modern forecasts impact everything from the economy, to public safety, travel, agriculture, and even the military. In the US the National Oceanic and Atmospheric Administration, related federally funded institutions, and a multitude of universities, together with similar organizations in other developed countries (particularly Europe) have collectively made great strides in developing forecasts that are applicable to nearly all points on the globe. Inconceivable only 20 years ago, it is possible now to download to a cell phone 5-day weather forecasts for thousands of destinations around the world with just a few button clicks.

With such a revolution in weather forecasting, what benefits has astronomy derived from such forecasts in the late 20th century? One might think they would be enormous, but in reality, astronomers gained relatively little in observing efficiency from these forecasts. They no longer had to be amateur meteorologists to develop their own forecasts, and satellite images made it much easier to gauge the extent of cloudy periods. Still, these forecasts, which were typically developed for public use, lacked detail in key areas that benefit modern astronomy facilities that are capable of *adapting* to changing weather conditions. A case in point is how weather forecasts were disseminated to astronomers at Mauna Kea and used by them only 15+ years ago. The typical routine of astronomers there (and Mauna Kea is the norm, not the exception) is to gather in the common eating facility in the late afternoon, discuss the night's plans and/or last night's results with colleagues using other observatories, and then drive up to the summit to begin a long night's work. Before leaving though, would be the "traditional" hunt for a recent copy of the local newspaper. Therein would hopefully be a tiny gray-scale image of the central Pacific that would give some hint of the cloud field surrounding the Hawaiian archipelago, together with basic forecast parameters like the high and low temperatures of various small towns on the Big Island. That was it. On a good night it was possible to find a copy of the newspaper that was less than 1 day old – on a bad night only the classified ads could be found. The link to the nearest forecast

Figure 7.3 Techniques used for centuries by astronomers to get the "upper hand" on the weather included such tools as the barometer and almanacs. These tools, combined with an ability to read cloud forms, wind directions, and temperature changes allowed astronomers for centuries to make an informed guess about what the weather would be each night.

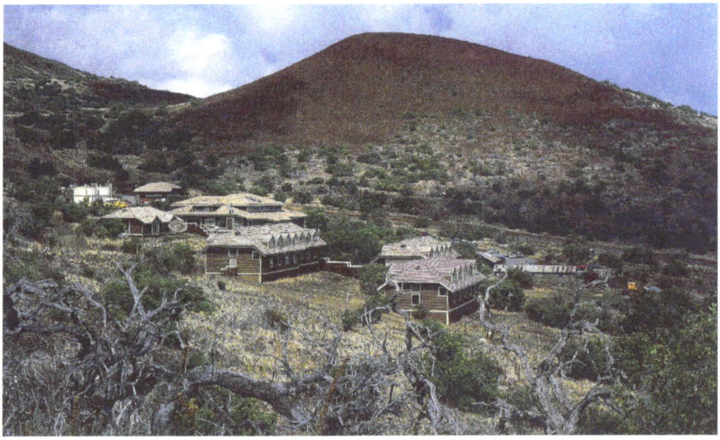

Figure 7.4 Adapted from West (2005), the mid-level facility known as Hale Pohaku is the evening gathering place for astronomers who come from around the world to observe from the summit of Mauna Kea.

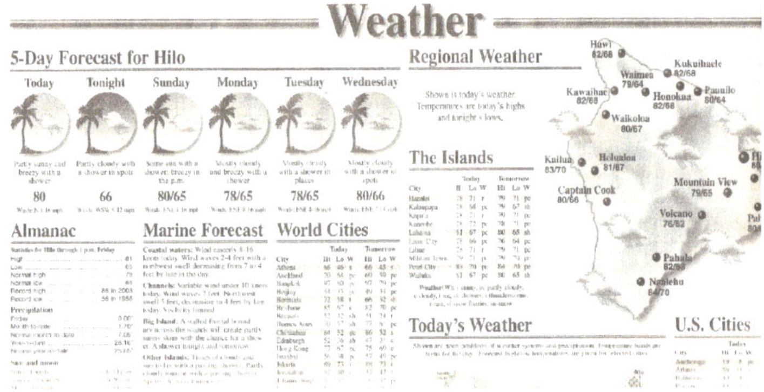

Weather

5-Day Forecast for Hilo

Today	Tonight	Sunday	Monday	Tuesday	Wednesday
80	66	80/65	78/65	78/65	80/66

Almanac

Marine Forecast

World Cities

Regional Weather

Shown is today's weather. Temperatures are today's highs and tonight's lows.

The Islands

Today's Weather

U.S. Cities

Figure 7.5 Adapted from the Hilo *Tribune Herald*, this level of detail was the norm for astronomy forecasts on Mauna Kea until the advent of the Mauna Kea Weather Center.

office was "broken" in both directions though, as it was not at all unusual to have the National Weather Service forecast office in Honolulu call observatories in the middle of the night to learn how high the wind speed is (in case a high-wind advisory needed to be posted) or if it was snowing (in case a winter storm advisory is needed). Finally, all observatories at the time were using so-called "classical" techniques, in which astronomers would apply for telescope time well in advance of when it is needed, have their proposals peer-reviewed, and if awarded time, it would be scheduled months in advance. After traveling thousands of miles, a typical observing run would last 2-3 nights and—as mentioned before—the success of the run could be dominated by the weather. Given the millions of dollars spent each year to run these advanced facilities, using "classical" observing techniques that were developed long before weather forecasting existed, change was needed to enhance the scientific product and cost effectiveness of modern observatories. On Mauna Kea alone nearly USD700,000,000 has been invested to date and about USD90,000,000 is spent each year to maintain and operate all of the facilities on Mauna Kea. With over 500 people employed by astronomy on the Big Island, many offering interesting high-tech positions with good salaries, the economic impact of the astronomy "industry" is estimated to be ~USD200,000,000 annually. For an economy the size of the Big Island's, these are important factors, but arguably even more important is the science impact on humanity, stemming from discoveries made on the summit of Mauna Kea. To date an astonishing ~6,000 publications have been based upon observations from Mauna Kea. It is difficult to put a dollar value on what these observations have contributed to society at large, but it is safe to say that our understanding of the universe would be *substantially* diminished if this enormous knowledge and discovery base were never created. For cold hard economic reasons or more esoteric but lasting and meaningful scientific reasons, the techniques used to perform research were about to change by the turn of the 21st century and, at last, the "missing link" between advanced forecast technologies and the world's greatest collection of telescopes would be found in the form of the Mauna Kea Weather Center.

7.2 The Power of Queue Observing From the Ground

The time-tested classical observing technique that has been used for centuries has some advantages but, from a broader strategic perspective, has some real shortcomings as well. It enables serendipitous discoveries when the astronomer most familiar with targets being observed is at the "helm" and with expertise is more likely to be able to recognize and adapt to new discoveries than is an astronomer taking data on another's behalf. The "helm astronomer" also understands best the nuances of the observations, including when the data is truly "good enough" to support the research program in terms of signal-to-noise in recorded images or in spectra and image quality. Also, classical observing helps to maintain an important bond between guest astronomers and the staff responsible for maintaining and operating the observatory. A good relationship and mutual understanding between any vendor (observatory) and customer (astronomer) benefits both parties over the long term. It also serves to train the next generation of young astronomers in the process of actually using these rather complex systems—something that really requires in-situ training to fully comprehend.

While classical observing is fine for observatories that have fixed configurations (e.g., one instrument available at a time) or very few astronomers in their user base, it is not well adapted to major national facilities required to meet the needs of thousands of astronomers, with a broad science mission, and considerable flexibility to adapt to changing conditions. The current generation of observatories is also much better at exploiting local site conditions than previous-generation observatories, particularly in the area of image quality (seeing). The importance of temperature-controlling telescope structures and flushing air through domes is now taken for granted, when only a couple of decades ago it was still under investigation as a possible performance inhibitor. Modern site-limited telescopes are now more subject to the vagaries of conditions than self-limited telescopes of the past, meaning that the quality of the data they produce will have greater extremes. This forces modern observatories more and more to queue-based systems that offer much greater flexibility to accommodate changing conditions.

In the early days of the Gemini 8-m Telescope Project, a number of models were considered to operate these exciting new facilities. These telescopes were unique in that they are highly integrated systems, with the telescope, enclosure, instruments, and data systems all falling under central computer control so that observations can be scripted and executed very efficiently. In addition Gemini has 3 ports available all the time for instruments, plus an adaptive optics unit and a calibration system that together allow observations from optical through the mid-infrared at anytime. As a result, observations that can be carried out through thin cirrus, or in times of poor seeing or of extremely low water vapor (e.g., at 20 μm), can be executed in series and over many nights until entire observing programs are completed as conditions change. This is a completely different approach to running an observatory and—initially—considerable skepticism was expressed about its viability. Can staff astronomers really acquire data as high in quality as that possible as visiting PI's? Can the open-shutter efficiency be as high as classical observing when telescope and instrument configurations are changing all night? These and many other questions were posed in the early days of choosing which operating model makes the most sense for an observatory like Gemini. In an attempt to understand the effectiveness of classical vs. queue-based operations, a number of simulations were run to see how well observing proposals could be matched to changing conditions. The underlying operating principles and results of these queue simulations were captured in a document (Fig. 7.6) that demonstrated unequivocally that a queue-based system is better able to execute the highest scientific ranked proposals than is a classical system. If an overarching metric for running a observatory like Gemini is to maximize its scientific product, then a change in culture would be needed to lead the community toward a better approach to acquiring data for astronomers.

With the Gemini telescopes in place and operational around 2000 (Gemini-North) and 2002 (Gemini-South), and the decision made to proceed with queue operations at the 50% level (the other 50% being classical observing), Gemini needed to find a much better way to acquire weather forecasts than had ever been

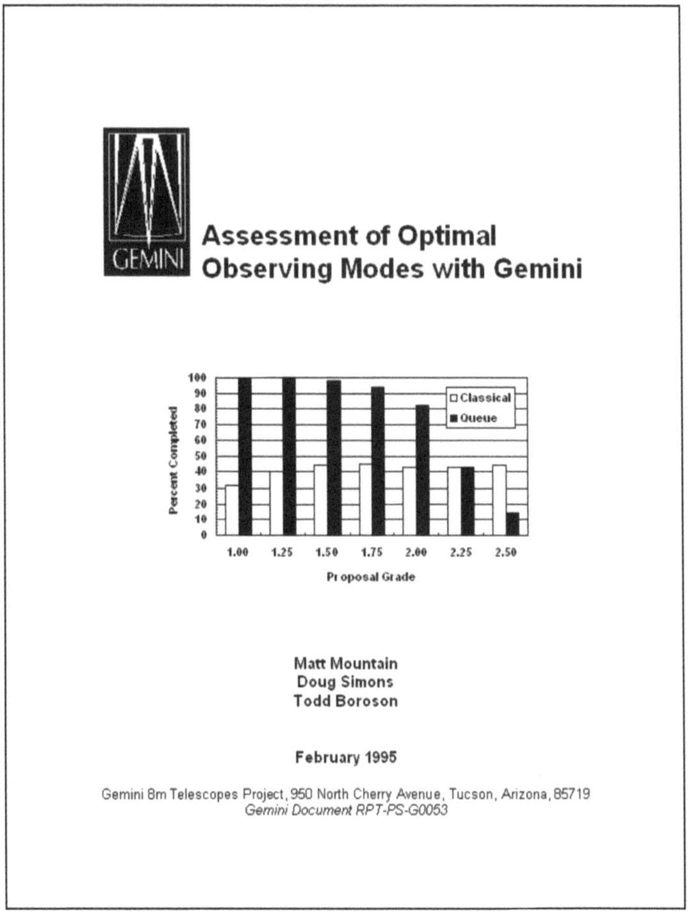

Figure 7.6 A seminal document released by Gemini in 1995 assessed the merits of adopting a completely new observing model—called queue observing.

achieved in the history of astronomy. Weather forecasts meant the difference between operating reactively vs. proactively in the execution of the queue, and from a tactical perspective were indispensable. Acquiring such forecasts required exploration of several possibilities but, as explained below, a certain amount of good fortune made all the difference.

7.3 Astronomer Meets Meteorologist

Possible sources of weather forecasts initially considered included a completely automated service that would provide regular–albeit rather minimal–forecasts via the internet to seeking help from the Honolulu NWS office to issue custom forecasts specifically for the summit of Mauna Kea. Through the Gemini Board of Directors member based at the University of Hawaii (Bob McLaren) a meeting was organized between McLaren, Douglas Simons, and the head of the NWS in Honolulu to discuss possibilities. Making the case to the NWS that such forecasts were justifiable given their finite resources and the need to issue other custom forecasts that supported public safety, fire conditions, etc. was not going to be easy. The results of that meeting were captured in an e-mail message sent from Simons on 11 November 1996 to then Gemini Project Scientist Fred Gillett, the essence of which is quoted below.

"Bob [McLaren] and I spent a couple of hours down on the UH campus discussing ways to foster a closer relationship between Big Island astronomy and weather forecast products generated in Honolulu. I started out rather skeptical about the way the meeting was proceeding—basically just getting a core dump on various standard products the NWS churns out and I wasn't learning anything new, but during a daily debriefing of the NWS team we met a UH meteorologist (Steve Businger) and Bob and I rapidly ditched the NWS "tour" to talk to Steve off-line about forecasting needs for astronomy..."

This chance encounter between McLaren, Simons, and Businger was the genesis of what has come to be known the Mauna Kea Weather Center (MKWC). With experience in the use of MM5 and a keen interest in exploring new possibilities of developing targeted high-resolution forecasts, the needs of Gemini-North (and in time essentially all Mauna Kea Observatories) and the interests of a key faculty member at the University of Hawaii's meteorology department meshed nicely. Still, the resources and infrastructure needed to provide regular forecasts did not exist and considerable ground work would be needed to turn the collective vision of a few scientists into a functioning forecast system. The necessary resources included the following:

- Financing shared across all of the Mauna Kea Observatories
- New computers to interpret models and post forecasts
- A couple of meteorologists to generate daily forecasts, to maintain the considerable data network needed to generate forecasts, and to develop radically new forecast products like seeing
- Access to a super-computer to generate the complex 3D high-resolution forecast models at least 72 hours into the future
- Organization across the Mauna Kea Observatories to provide guidance on what forecast products are really needed and to prioritize the development of new products
- A reliable source of data (satellite, balloon, PWV, etc.) to ingest into these high-resolution models

- Overall management structure to hire new staff, procure equipment, provide travel funding between Honolulu and the Big Island

Going from a well-founded vision to a working forecast office was a tall order, but synergies were identified across many entities that, in the end, led to a remarkable forecasting operation. Figure 7.7 shows the key elements involved, which include a Fujitsu supercomputer that Subaru Observatory—another very young 8-m facility on Mauna Kea— generously provided for the purposes of running MM5 on a daily basis. This contribution alone would have been prohibitively expensive to purchase or rent at the time and allowed the MKWC to fast-track its development program, giving it credibility in the delicate early years when their "customers" were a bit wary of the reliability of these new forecast products. More than just a hardware contribution, Subaru provided the technical expertise needed to optimize the execution of MM5 code on this platform. Luckily, demand for computing time on this supercomputer by Subaru astronomers for data processing was low enough that the MKWC had few problems consuming large fractions of the computing power of this machine, meaning MM5 could be used operationally, not just as a development tool. The University of Hawaii's Institute for Astronomy acted as the resource catalyst in the endeavor. Seed funding for the program was routed by the IfA to the MKWC to purchase new computers and hire a meteorologist to issue daily forecasts. In time the IfA's role would expand to help galvanize support for the MKWC in its early days with the Mauna Kea observatories who knew that once the seed money was used they would be asked for annual contributions to support the MKWC financially on a steady basis. The UH Department of Meteorology provided the core expertise needed to issue forecasts and a long-term vision for how the MKWC would develop in the years ahead. Using existing links to the National Weather Service forecast office in Honolulu, which is conveniently located on the UH campus, links to data providers were established that would be ingested into the forecasts generated in Hilo via the Subaru supercomputer. At the nexus of all of this activity, the UH meteorology department worked closely with the UH-IfA, Subaru, and Gemini to build all the elements needed to release core forecast products like summit temperature, PWV, and cloud cover in the early days of the MKWC. Soon a group called the Site Monitoring Working Group (SMOWG) would be established, with representatives from all of the Mauna Kea observatories, that would act is the "User Committee" for the MKWC, providing feedback on the value of forecast products and prioritization for their development over time. Though the SMOWG meets only twice a year, it provides a natural and direct link between all of the Mauna Kea observatories and the MKWC staff and serves as a mechanism to procure new summit monitoring equipment using contributions from all observatories.

Each component or "input" in Fig. 7.7 needed to come together through the MKWC in its vulnerable early days. As such, this represents a remarkable synergy

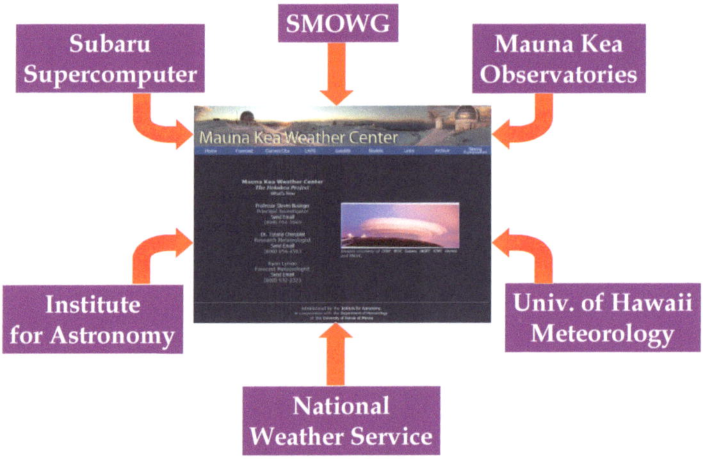

Figure 7.7 All of the elements that had to come together to create and support the Mauna Kea Weather Center are shown. This is a remarkable assemblage of resources all focused on providing forecasts for astronomy–unlike anything available in the entire history of astronomy.

of needs, resources, and expertise that merged to create something totally new in astronomy. Together they represent planks in the foundation of a sophisticated forecasting system that is now used daily to critically support operations of the world's largest and most advanced array of telescopes high atop Mauna Kea. Naturally, astronomers tend to receive notoriety when discoveries made from the Mauna Kea observatories are released to the public. Little do they know that those discoveries are being "quietly" and crucially backed up by weather forecasts unlike any provided in the history of astronomy.

7.4 Putting it all Together at Gemini Observatory

Each day at Gemini Observatory an astronomer who has been assigned the role of "Queue Coordinator" (QC) prepares an observing program for the next night. This program consists of portions of pre-scripted observations that astronomers world-wide have submitted to Gemini. These control the telescope, dome, and instrumentation in a carefully optimized sequence to ensure maximum observing efficiency. A wide variety of parameters are taken under consideration by the QC as nightly plans are developed, including scientific ranking of the proposals, the extent to which proposals have already been completed, national time shares, the instrument configuration for the night, and, of course, the forecast weather. Sophisticated planning tools developed at Gemini are used to select from all of the proposals submitted that semester and give the QC instant access to a myriad of information stored in Gemini's central observing database. As objects rise and

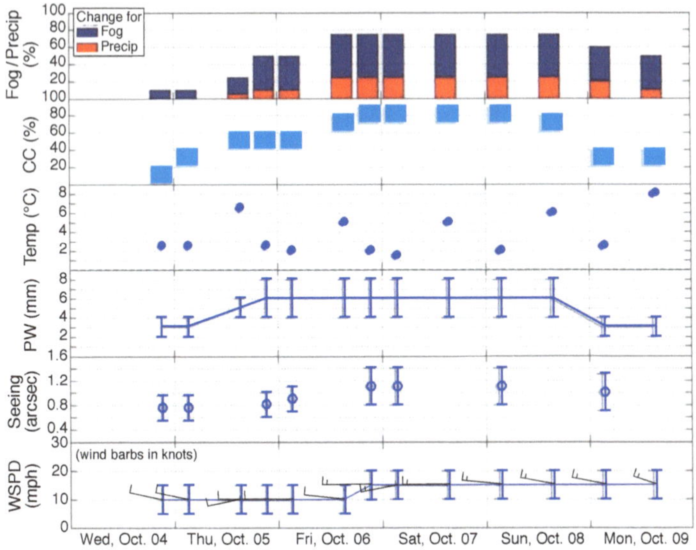

Figure 7.8 The forecast products issued twice daily from the MKWC include a diverse range of variables specific to the summit of Mauna Kea. Those used most frequently are shown graphically above and include the chances for fog/precipitation, clouds, summit temperature, PWV, seeing, and wind speed/direction. At last, in one glance astronomers can determine what the weather will be over the course of their observing run with confidence. Compared to Figure 5 it is easy to see why this truly represented a quantum leap in forecasting systems. Such forecasts are totally unique in the history of astronomy.

```
GN NightLog: 2007 mar17-mar18
Submitted by : T.Dall
              Total hours lost to faults           : 0
              Total hours lost to weather          : 0
              Total hours observed on sky          : 10.33
              Time between nautical twilights      : 10.33
     Science Programs Observed: GN-2007A-Q-51, GN-2007A-Q-49, GN-2007A-Q-33, GN-2007A-DD-4,
   GN-2006A-Q-8, GN-2007A-Q-8, GN-2006B-Q-18, GN-2007A-Q-10, GN-2007A-Q-81, GN-2007A-Q-12

     Engineering Tasks Observed: ENT117

     SUPPORT:
     Summit Observers   : T.Dall, B.Walls
     Base Facility      : R.Mason
     Phone/Polycom      :
     INSTRUMENT(S) Used : GMOS-N NIRI MICHELLE


     TELESCOPE USAGE:
        Science                                 : 10.25
        Commissioning                           : 0
        Engineering                             : 0.08
        Telescope Shutdown                      : 0
```

Figure 7.9 The top of a typical night log at Gemini-North is shown. In the "Science Programs Observed" line are the program ID's for the various elements of the queue executed that night. In this case portions of 10 separate programs were observed using 3 instruments, offering optical (GMOS-N), near infrared (NIRI), and mid-infrared (MICHELLE) capabilities.

set over the course of the night, the entire observatory functions under central computer control through a choreography that is crafted by the QC to make it as simple as possible for the night staff to operate all of the telescope systems on the summit, where lack of oxygen and fatigue make it challenging to function well. Of course, an overarching constraint the QC's use to generate nightly plans is that observing programs should be selected on the basis of the forecast weather conditions. For example, if cirrus is forecast, then perhaps only GMOS pre-imaging will be done in advance of future multi-slit spectroscopy observations, or spectroscopy of relatively bright targets will be recorded if it does not need to be photometrically calibrated to great precision. Likewise, if a particularly dry night is forecast, the corresponding nightly plan will be biased toward inclusion of mid-infrared programs using MICHELLE.

Over the course of each night, a report or "night log" is generated by the staff astronomer executing the nightly observing plan, which is sent to the entire observatory in the morning to allow the day staff, including a range of engineers, technicians, administrators, and scientists, to provide various support roles in advance of the next night's observations. Figure 7.9 shows an example of such a night log. Over the 10.33 hours of observing time used that night, observations were made for parts of 10 different science programs using optical, near-infrared, and mid-infrared instrumentation. Though the astronomer running the queue that night was able to adjust the plan over the course of a night, the plan is initialized on the basis of the morning forecast. Changes in the plan may occur due to weather conditions changing unexpectedly (a poor forecast), or problems with the instrumentation or telescope that force the use of only certain capabilities or modes, or the arrival of a target-of-opportunity alarm signaling that the telescope needs to be quickly pointed to a new target that is changing rapidly (e.g., a supernova or gamma ray burster). Whatever the reason, Gemini's queue-based system is flexible enough to accommodate changes to ensure that regardless of the current situation, the telescope is being used wisely to gather as much science data as practical. In essence, Gemini's queue "closes the loop" between the MKWC forecasts and world-class research.

7.5 Conclusions

Unlike nearly all other fields of science, astronomy is uniquely subject to the vagaries of the skies, and for centuries uncertainties in weather conditions have impacted astronomical research. While we clearly cannot change the weather, being able to predict and adapt to changing weather conditions gives modern astronomers the ability to intelligently work around the weather. This has a significant impact on such world-class sites as Mauna Kea, Hawaii, where optical, infrared, and radio observatories function constantly amid a dynamic climate. The MKWC arose not out of just need, but from a unique synthesis of resources, skills, and vision to address this problem in ways never before accomplished in astronomy.

Today, high-resolution weather forecasts that include such crucial parameters as cloud cover, PWV, and even seeing, are issued by the MKWC and used regularly to guide astronomy research that substantially defines the limits of civilization's understanding of the universe and our place in it. In the diverse global arena that astronomy functions in today, the MKWC is not widely known. In time though, it will come to be known as the vanguard of astronomy forecasting systems, and we astronomers will all wonder how previous generations ever got along without it.

Acknowledgments

The Gemini Observatory is operated by the Association of Universities for Research in Astronomy, Inc., under a cooperative agreement with the NSF on behalf of the Gemini partnership: the National Science Foundation (United States), the Science and Technology Facilities Council (United Kingdom), the National Research Council (Canada), CONICYT (Chile), the Australian Research Council (Australia), CNPq (Brazil), and Ministerio de Ciencia, Tecnología e Innovación Productiva—Presidencia de la Nación (Argentina).

Authored by
D. A. Simons
Gemini Observatory, Northern Operations Center, Hilo HI

J.-R. Roy
Gemini Observatory, Southern Operations Center,
La Serena, Chile

Optical Turbulence Modeling and Forecast— Towards a New Era for Ground-Based Astronomy

Optical turbulence characterization for astronomical applications done with atmospheric mesoscale models is a relatively recent discipline. First studies appeared in the nineties. Simulations retrieved from these models can be fundamental for optimization of Adaptive Optics (AO) techniques and characterization and selection of astronomical sites. In most cases, simulations and measurements can provide complementary information on turbulence features. The synergic employment of both approaches provides a better understanding of the fundamentals of the nature of turbulence and characterization of its spatial and temporal evolution.

The potential of the numerical approach and the fundamental scientific challenges related to mesoscale atmospheric models rely upon the ability to (1) to describe a 3D map of the C_N^2 (i.e., the volumetric distribution of the optical turbulence in the troposphere) in a region around a telescope, (2) forecast the optical turbulence (i.e., know in advance the state of the turbulence conditions above an astronomical site), and (3) perform a climatology of the optical turbulence extended over time scales of the order of decades. No other comparable tools of investigation are available that can achieve these three scientific goals. However, these highly challenging goals are associated with an intrinsic difficulty in parameterizing a physical process, such as turbulence, evolving at spatial and temporal scales smaller than those usually explicitly resolved by a mesoscale model.

In this paper the most fundamental challenges for the simulations of optical turbulence for astronomical applications will be discussed. The methods usually employed to model and forecast optical turbulence in astronomy will be presented beginning with statistical approaches and neural networks, analytical and physical models, and concluding with numerical models. We will describe what a mesoscale atmospheric model is and how the dynamic and optical turbulence are described in such a model. The main results and progress achieved so far in this field and the most important challenges for the near and far future research will be summarized. We will conclude with a brief presentation of the main research lines and motivations supporting the ForOT project.[1]

---

[1] ForOT: '3D Optical Turbulence Forecast above astronomical sites', http://forot.arcetri.astro.it

8.1 Modeling Optical Turbulence for
Astronomical Applications—Challenges

Apart from the pioneering statement announced by Newton (Optika, 1963)[2], the idea that site selection needs to be done according to scientific criteria has been affirmed only recently (1962) when the International Astronomical Union (IAU) set up a commission[3] for the identification and protection of astronomical sites. Since that time, several sites for 8-10 meter class telescopes have been identified after intense site testing campaigns aiming to quantify the turbulence typically developed during the night time on the summit of mountains, preferably located in dry and remote regions of the Earth. In 1998, the European Southern Observatory (ESO) created the European Search for Potential Astronomical Site (ESPAS) working group to support the site selection for the new generation ground-based telescopes, the so-called Extremely Large Telescopes (ELT) [pupil size D $\geq$ 30 m]. Concurrently, the Thirty Meter Telescope (TMT) and the Giant Magellan Telescope (GMT) projects (the American alter ego of the European E-ELT) set up working groups with similar goals. Site selection science had therefore taken, in the last decades, a fundamental role in ground-based astronomy.

At present, we can state that the future of ground-based astronomy relies upon the potential and feasibility of the new class of ELTs. In spite of the presence of the atmospherical turbulence, the ground-based observations have, with respect to the space-based observations, at least a few fundamental advantages that justify the continuing interest of astronomers in the ground-based facilities,

(1) lower financial investment
(2) longer typical lifetime of facilities
(3) relative ease of maintenance and upgrade
(4) better angular resolution achievable with typically larger pupil sizes.

This can be obtained by the use of Adaptive Optics (AO) techniques (Roddier, 1999), expressly to reduce/eliminate perturbations produced by the atmospheric turbulence on the wavefronts coming from the stars. For space-based observations, spatial resolution is proportional to pupil size. However, without AO techniques the atmospheric turbulence in ground-based observations inexorably limits the angular resolution of images to that obtained with a small 10 cm telescope. To correct the turbulence effect and preserve the potential of ground-based astronomy we need to know the nature of the turbulence itself. For this reason the characterization of optical turbulence (C_N^2 profiles and integrated values) and, more generally, studies aiming to better define mechanisms of formation and development of the turbulence, are crucial elements to guarantee successful ground-based observations and feasibility of ELTs. Today astronomers need to know how much turbulence is developed above an observatory and how the turbulence is distributed in the troposphere (~20 km).

---

[2] From Optika: 'The only remedy to poor telescope performance is almost serene and quiet air, such as may be found a top the highest mountains, above the grosser clouds'.
[3] Commission of Site Identification and Protection

These studies support the selection and scheduling of scientific programs to be realized with different instruments placed at the telescopes foci (flexible-scheduling), selection of the best sites for new telescopes, and employment of the most recent AO techniques such as the Multi-Conjugated Adaptive Optics (MCAO) (Beckers, 1988; Tallon and Foy, 1990; Ragazzoni et al., 2000), Laser Guide Star (LGS) (Foy and Laberie, 1985; Thompson and Gardner, 1987), and Ground Layer Adaptive Optics (GLAO) systems (Rigaut et al., 2000). Despite the fact that all projects of new ELTs plan to have a flexible-scheduling system to optimize the management of telescopes, currently only a few attempts at real flexible-scheduling systems based on the forecast of optical turbulence exist but they are at an embryonic stage. At the same time, an efficient search for and selection of the best astronomical sites in the world is difficult because measurements are not homogeneous, site testing campaigns are expensive, and time required to characterize a site with measurements can extend to decades.

So far, a few instruments able to measure one or several different astroclimatic parameters characterizing the wave-front perturbations induced by the atmospheric turbulence have been conceived, built, and employed. However, all these instruments have a common limitation: they provide local measurements. Measurements can be taken in situ, as in the case of radio soundings and instrumented masts (Azouit and Vernin, 2005), or along a single line of sight, as in the case of optical instruments such as the Generalized Scidar (GS) (Fuchs et al., 1998; Avila et al., 1997; Klueckers et al., 1998) or the Multi Apertures Scintillation Sensor (MASS) (Tokovinin et al., 2003). What is not accessible by such measurements and can be offered by the numerical technique are 3D maps of the turbulence around a telescope and 3D maps of the turbulence predicted in the future. This fundamental information can be obtained with non-hydrostatic mesoscale atmospherical models. Our studies using the Meso-Nh[4] model contributed in a consistent way to the development of the numerical discipline in an astronomical context from our first attempts (Masciadri et al., 1999a,b) up to the most recent and reliable results (Masciadri et al., 2004; Masciadri and Egner, 2006). The potential of this numerical discipline supports further scientific work in this direction. The main advantages achievable with the numerical technique are:

(1) construct 3D C_N^2 maps above a surface region of a few tens of kilometers around a telescope;

(2) simulate the C_N^2 'where' and 'when' it is required without the necessity of long and expensive site testing campaigns done with several instruments; this is particularly interesting for searches of new sites;

---

[4] Meso-Nh has been developed by the Centre National des Recherches Meteorologiques, Meteo France and Laboratoire d'Aerologie, Université P. Sabatiér, Toulouse, France. See Sec.3 for detailed information.

(3) forecast C_N^2 profiles, (a real 'chimera' for all astronomers), fundamental information for the flexible scheduling of the scientific programs and instrumentation;

(4) set up a climatology of the optical turbulence extended on a time scale of the order of decades. It is worth remembering that only models (and not measurements) can access the past, and this condition is required to reconstruct a climatology on long-term time trends and better predict site quality. Climatology studies of the optical turbulence are necessary, for example, to quantify effects evident on time scales of decades, such as deterioration of the seeing already observed at the summit of a few astronomic sites. The solution of this problem is one of the most challenging goals for site testers at present. The most known and famous case is probably Cerro Paranal (Chile), site of the VLT. Figure 8.1 shows the seeing at Cerro Paranal during the nineties when the site was selected for the VLT. After a decade (~2003), the seeing appears visibly worse as shown in the same picture.

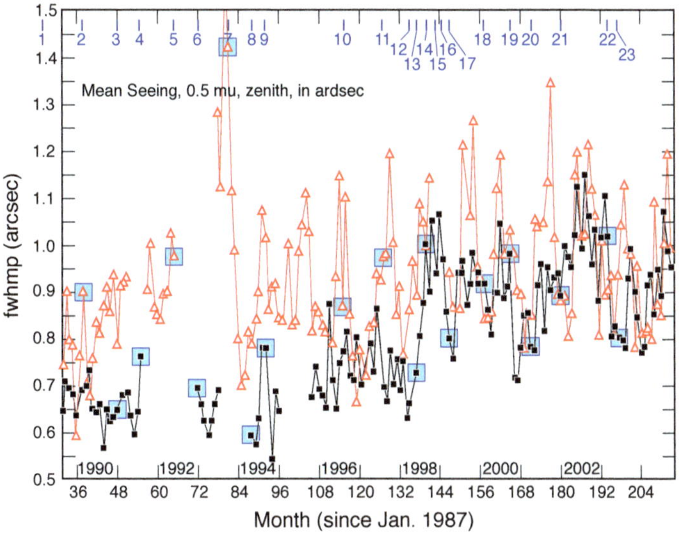

Figure 8.1 Mean seeing values measured from 1990 to 2003 above Cerro Paranal (black line) and La Silla (red line) sites (from ESO Web Site).

No other tools of investigation with comparable potential exist at present to achieve these four scientific goals.

Astronomical observations related to the most challenging scientific programs frequently require excellent turbulence conditions. Our ability to forecast these particular conditions determines our progress in astrophysics knowledge and gets ground-based astronomy competitive with space-based astronomy. The scientific impact obtained with flexible scheduling based on the forecast of the optical turbulence is quite remarkable. Roddier and Lena (1984) showed that, for a long base-line interferometer with two large telescopes, the signal-to-noise ratio varies at least as r_0^3 and that by increasing r_0 from 10 cm to 20 cm (equivalent of an improvement in seeing from 1 arcsec to 0.5 arcsec) it is possible to gain a factor of 2 in the magnitude limit of the detectable scientific objects. It is possible to show (Veron-Cetty and Veron, 1996) (Fig. 8.2) that passing the magnitude limit of 12 to 14 in the visible, the number of Active Galactic Nuclei (AGN) detectable from the ground and with a declination between -70° and +10°, increases from 5 to 100. This simple example shows, in a very impressive way, that the potentialities obtained with a predictive system of turbulence conditions should have an extremely important scientific impact on ground-based astronomy.

But what is the nature of light/matter interaction? Turbulence stochastically perturbs the wavefront (i.e., an electromagnetic field E(x,y)) and modifies its amplitude A(x,y) and phase ϕ(x,y). When a perfectly plane wavefront propagates through the atmospheric turbulence, fluctuations of temperature ΔT induce fluctuations of the refractive index Δn that finally perturb the parameters characterizing the wavefront: A(x,y), ϕ(x,y). All observational astronomy methods, such as direct imaging, spectroscopy, and interferometry, are based on the measurement of signals related to A(x,y), ϕ(x,y) or both. In the sixties, Tatarski (1961) in Russia and Fried (1965, 1966) in the United States contributed by elaborating the theory of interaction between wavefronts (light) and turbulence

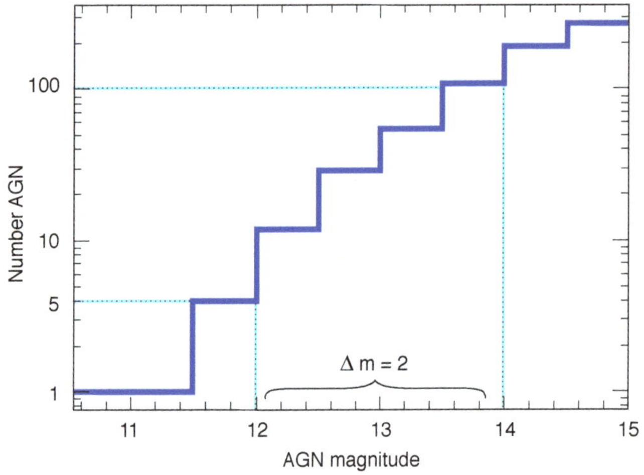

Figure 8.2 Cumulative histogram of AGN magnitudes V ($V \leq 15$) for a declination in the [−70, +10] degrees range (Véron-Cetty and Véron, 1996).

(matter) i.e., defining the relationship between A(x,y) and ϕ(x,y) with parameters characterizing the macroscopic state of the atmosphere (fluctuations of pressure, temperature, wind speed, humidity and refraction index). More recently, Roddier (1981) reformulated such a theory for astronomical applications.

8.2 Methods to Forecast the Optical Turbulence

In this chapter, methods and results obtained so far in forecasting optical turbulence using different approaches are reviewed. This presentation was inspired by Masciadri (1998) but we extend the analysis here to recent results. Particular attention is given to the advantages and disadvantages of each technique and the main potentialities intrinsically related to the numerical approach: non-hydrostatic mesoscale atmospheric models, the main topic of this paper.

8.2.1. Statistical Method: Nowcast

A multiregression technique (called *nearest neighbor regression*) has been applied in the past to the astronomical sites Paranal and La Silla (Murtagh and Sarazin, 1993) with the aim to predict information on the seeing[5] (ε) and the temperature in proximity of the ground. This was not exactly a forecast but a nowcast aiming to relate the seeing to the meteorological and environmental conditions observed at the same time or in the near past. This technique gave interesting results for the temperature forecasted at 24 hours at 2 m and it showed a 62% success rate within 0.5 K. The success rate increases to 85% if we relax the precision within 2 K. What are the implications of this result? Racine et al. (1991) calculated experimentally that the seeing depends on the difference of temperature between the primary mirror and the surrounding air as $\varepsilon_m = \alpha_m \cdot \Delta T_m^{6/5}$. With α_m = 0.4 (value calculated by the authors for the Canada-France-Hawaii Telescope [CFHT]), we can argue that a ΔT_m = 2 K can easily trigger a seeing of roughly 0.9 arcsec in the visible. This means that the prediction of the temperature with a precision better than 2 K is suitable to guarantee sub-arcsecond turbulence conditions. The main limitation of such a method is that it cannot predict sharp changes in temperature from 2 K up to 6 K within 24 hours, an event that occurs 15% of the time in the season studied (summer 89-90). The results on seeing (ε) by Murtagh and Sarazin (1993) were encouraging but they had some limitations. For example, with regard to the categories of good and bad seeing with respect to the average, the authors found that the values of some particular combinations of parameters (pressure, wind speed and direction, temperature, humidity) associated with good seeing is observed in 93% of the cases but, unfortunately, the probability of obtaining these combinations is only 10%. If we consider the probability of being able to do a prediction, i.e., the number of the possible set of

---

[5] The word "seeing" is used in astronomy to indicate the integrated value of the optical turbulence along the whole troposphere.

values for a given set of parameters (we define "set of values" those that occur at least four times), we find that the probability of obtaining a good seeing is 12.5% (spring), 8.9% (summer), 20.4% (autumn), and 32.2% (winter).

8.2.2 Neural Networks

The artificial neural network is a mathematical structure able to identify complex and non linear connections between elements of several data-set. The interaction reaches an autoadaptation status achieved in a dynamic and interactive way. A particular neural network has been tested for the forecast of the seeing at Cerro Paranal (Chile) by Aussem et al. (1994). The advantage of the artificial neural network with respect to the statistical methods is that the former have a sort of internal memory permitting it to re-assimilate recent and past information in a sort of continuous feedback. This potential advantage proved futile in that study due to an important temporal discontinuity of data (temporal seeing values flowchart). The technique gave less encouraging results than those from the nearest neighbor regression. A further estimate of the reliability of the artificial neuron network method has been given by Buffa and Porceddu (1997): the temperature near the ground has been predicted at 12 hours with an error of 2 K and at 6 hours with an error of 1 K.

8.2.3 Analytical Models

Several attempts were made to describe analytically the vertical distribution of the optical turbulence (C_N^2 profiles). The most well known is probably the power-law Hufnagel model (Hufnagel, 1974), adapted (in proximity of the ground) for unstable day time conditions ($\sim h^{-2/3}$) as well as stable night time conditions ($\sim h^{-4/3}$) (Wyngaard et al., 1971b). The Hufnagel model has proved to be roughly compatible with measurements (Barletti et al., 1976). The main principle on which the analytical method is based on is a fit done on measurements and aiming to define average $< C_N^2 >$ profiles representing the typical optical turbulence vertical distribution. In the literature we can find some small differences, principally based on the typology of the fit. The Hufnagel model is mainly based on Bufton's (1973a,b) measurements.

More recently, Trinquet and Vernin (2006) provided evidence of the existence of thin sheets of strong thermal shears in the troposphere and stratosphere and they proposed a model where C_N^2 depends on both the average $< C_N^2 >$ and gradient of the potential temperature average $\partial \overline{\theta}/\partial z$:

$$C_N^2 = < C_N^2 > \cdot [A(h) \cdot (\partial \overline{\theta}/\partial z)]^{p(h)} \qquad (8.1)$$

The free parameters $A(h)$ and $p(h)$ are fixed in different slabs of troposphere and stratosphere by a fit calculated on a set of balloon measurements of C_N^2 profiles.

The authors showed that, in the free atmosphere, performances of this model are comparable to those of the model considering the simple average $< C_N^2 >$. Better performance has been observed near the surface where the gradient of the potential temperature is particularly strong in nightly stable regimes.

We note, however, that the C_N^2 depends on the gradient of the potential temperature; therefore the value of $< C_N^2 >$ already reflects the dependency on $\partial\bar{\theta}/\partial z$. It is therefore not clear why $< C_N^2 >$ and $\partial\bar{\theta}/\partial z$ should be considered independent variables. Moreover, the C_N^2 depends not only on $\partial\bar{\theta}/\partial z$ but also on some other macroscopic parameters such as the wind speed gradient and mixing length (or kinetic energy). If we express C_N^2 as a function of $\partial\bar{\theta}/\partial z$ and the mixing length L we have that $C_N^2 \sim (\partial\bar{\theta}/\partial z)^2$ (Tatarski, 1961); but if we express the C_N^2 as a function of $\partial\bar{\theta}/\partial z$ and the turbulent kinetic energy E we have that $C_N^2 \sim (\partial\bar{\theta}/\partial z)^{4/3}$ (Masciadri and Jabouille, 2001, Eq. 6). In spite of results found by Trinquet and Vernin (2006), it is not clear why this model (in which the $< C_N^2 >$ and the $\partial\bar{\theta}/\partial z$ are considered independent variables), should give better results than the simple $< C_N^2 >$. One of the main drawbacks of such a model is that it is not based on physical assumptions but is the result of a mathematical minimization depending arbitrarily on two parameters $(A(h), p(h))$.[6]

Finally, Racine (2005) set up an analytical model depending on a set of free parameters aiming to reproduce the seeing measured above the astronomical sites simply knowing the latitude and the altitude of the sites. The free parameters are fixed, with a fit of the analytical model to measurements done so far in different sites in the world.

8.2.4 Physical Models

Physical models are those models for which the relationship between the C_N^2 and the mean values assumed by the meteorologic parameters is based on physical laws. Tatarski (1961) states that the C_N^2 depends on the gradient of the potential temperature and the dynamic outer scale (L_0) as:

$$C_N^2 = C_{emp} \cdot L_0^{4/3} (\partial\theta/\partial z)^2 \tag{8.2}$$

---

[6] In promoting the model expressed by Eq. (1), Trinquet and Vernin (2006) state that this model is less complicated and faster than the atmospheric mesoscale model. Although we respect the alternative opinions of our colleagues, we do not share the same judgement of value in this respect. The speed of the second method (mesoscale models) is strictly connected to the configuration of the atmospheric model and the computing system used. It is well known that, with suitable computing systems for an operational application, the time required for a forecast is short enough to be employed in a realistic way. It is extremely dangerous to present the problem of turbulence modeling as a search for a less complicated or quicker method. It is better to consider each method for its intrinsic potentialities trying to get the best from each of them.

where C_{emp} is empirically measured. Under the assumption of an universal dynamic outer scale, the C_N^2 can be retrieved from Eq. 8.2. Coulman et al. (1988) and Abahamid et al. (2004) proposed a slightly different universal outer scale and calculated C_N^2 from Eq. 8.2 with C_{emp} = 2.8. Two weak points should be noted in this approach:

- The universal dynamic outer scale, which in principle should be a prime ingredient from which the C_N^2 is retrieved, is calculated using the C_N^2 itself. We are therefore in a closed loop in which the C_N^2 and the dynamic outer scale L_0 are not independent.

- As observed by Masciadri et al. (2001), the value of C_{emp} found in the literature is, in reality, measured in the surface layer, and different values in the range (1.6-3.2) have been calculated by different authors (Hill, 1978; Wyngaard et al., 1971a,b; Ottersten, 1969). The value of C_{emp} has in reality a range of variability. Tatarski (1961) proposed a more suitable C_{emp} = 2.8 $f(Ri)$ in the surface layer, where $f(Ri)$ is a function of the Richardson number expressing the thermodynamic stability of the atmosphere. Both Coulman et al. (1988) and Abahamid et al. (2004) extrapolated the value of 2.8 to the whole troposphere, but the validity of such a law is not assured in this region.

Van Zandt et al. (1978, 1981) proposed a different approach—not to calculate a C_N^2 profile but to find a correlation between the C_N^2 and the Richardson number. The author introduces a probability function for triggering the optical turbulence. Comparing observations with calculations done under this assumption, the authors found that the C_N^2 peaks develop in proximity of the strong wind speed shears at the border of the jet-stream layer. Such a correlation has been successfully employed by Masciadri and Garfias (2001) and Geissler and Masciadri (2006).

8.2.5 Mesoscale Atmospheric Models

The numerical method is the only method based on a determinist description of all the physical parameters describing the temporal evolution of the atmospheric flow and its interaction with the ground. For this reason this method is the only one potentially able to identify and detect the physical sources triggering the turbulence and to reconstruct an abrupt change in the turbulence characteristics. In the next section such an approach is described in detail. Here the focus is on the challenge of using the numerical approach for the optical turbulence forecast. No other method can offer comparable perspectives, including statistical and analytical. Even in the context of the weather forecast, the numerical approach is the method the scientific community relies on. This is a concrete element supporting our choice in investing our efforts in developing the method of mesoscale atmospheric models to predict the optical turbulence.

8.3 Mesoscale Atmospheric Models

A numerical model is a representation of a real system (in our case the atmospheric flow) that aims to calculate the temporal evolution of the same. The model of a system is defined when we identify the variables necessary to describe it completely and unequivocally; when these variables are related to each other by an equal number of equations. In the case of the atmosphere we have seven variables: the three components of the wind speed $\vec{V}$, the pressure p, the temperature T, the density ρ and the water vapor concentration r. The equations are the conservation of momentum (three scalar equations), the first thermodynamic law (one scalar equation), the equation of the perfect gas (one scalar equation), the equation of conservation of the mass, and the water vapor concentration for a total of seven equations.

The physical principle of the atmospheric numerical forecast is deterministic. Knowing the state of the atmosphere at a time t_0, we can deduce the state of the atmosphere at each time $t > t_0$. In the case of forecasts made with atmospheric models, the initial state is a function of meteorologic observations done at a precise time in a finite number of meteorologic stations (radio soundings), aircraft measurements and, more recently, satellite measurements having a much larger coverage on the earth. This permits a more precise instantaneous picture of the initial condition. A numerical model predicts the temporal evolution of a set of variables calculated on fixed time steps (t_{step}). From a spatial point of view, the variables can be represented with the grid-points method or the spectral method. In the first case, each variable is calculated on a finite number of points (of a 3D network with finite size meshes). In the second case, the variables are represented by a linear combination of a set of orthonormal functions. For atmospheric models these are the circular functions of Legendre. In this paper we deal only with the first method. The size of the mesh that samples the variables gives us the resolution of the model. We therefore cannot access all physical phenomena developing at spatial scales smaller than the mesh size. In other words, all spatial frequencies higher than that defined by the mesh size are filtered out by the model. A fundamental concept in study of the dynamic of the atmosphere is the so-called "Scale Analysis". Each phenomenon is characterized by a typical spatial scale. A model is called explicit if the mesh size is smaller or equal to the typical spatial size of the phenomenon that the model intends to predict; and in this case, we can solve the temporal evolution equation of the variable describing this precise phenomenon. In the opposite case, we are forced to parameterize the variable, i.e., to represent the microscopic fluctuations of the variables as a function of the gradient of the same macroscopic variable, averaged on a large spatial scale. In this second case, the model is called implicit. We will refer later on to the Meso-Nh model, i.e., the non-hydrostatical atmospherical mesoscale model that we used and developed for our studies in astronomical applications.

The atmospheric models are called "mesoscale models" if the spatial mesh size is within the range of a few hundred meters (minimum extreme) up to 50-100 km (maximum extreme).[7] When the mesh size ΔX is smaller than the minimum extreme, models are called Large Eddies Simulation (LES); when the resolution ΔX is larger than the maximum extreme, models are called the General Circulation Models (GCM). These limits have to be considered as indicative, and it would be more appropriate to talk about order of magnitude than of precise values. The GCM are usually employed for weather forecasts, they extend over the whole earth and they describe phenomena having spatial and temporal scales of 10^2–10^4 km and 1–10 days respectively. Mesoscale models are used to describe phenomena characterized by smaller spatial and temporal typical scales such as orographic waves, turbulence, and deep convection. These phenomena (such as the turbulence) are not resolved explicitly but are parameterized in the model. This means that microscopic fluctuations of some parameters are expressed as a function of the same macroscopic variable averaged on a larger spatial scale. Such models need to be linked to the GCM to predict the parameterized variables. Mesoscale models are initialized at the time $t = t_0$ with analyses and/or forecasts provided by the GCMs. Initialization data are a sort of map of all the atmospheric variables extended on the whole earth and predicted with a low spatial and temporal resolution by the GCMs at the time $t = t_0$. The LES cannot be used for the forecast using the same initialization inputs describing the atmospheric flow status at a time t_0 on the whole earth because initialization products from the GCMs have a spatial resolution that is too low.[8] The mesoscale models represent therefore the ideal intermediate resolution size permitting access to temporal evolution of phenomena that are not resolvable explicitly.

In this context we are interested in one parameter reconstructed by a mesoscale model: turbulence. The classical turbulence model (Kolmogorov model) tells us that the isotropic turbulence is a phenomenon producing energy in the form of random fluctuations of wind speed of particles inside eddies of different sizes. The energy is injected at the level of the largest eddies (the dynamic outer scale L_0) and is transferred to smaller and smaller eddies to be destroyed by molecular and viscous dissipation at a small scale called inner scale (l_0), which is of the order of few millimeters in the atmosphere, but can reach a centimeter scale in extremely stable conditions (Masciadri and Vernin, 1997). The concept of the

---

[7] The subdivision in families can be more detailed. In this paper we do not distinguish between Limited Area Models (LAM) and mesoscale models to simplify the comprehension.

[8] Strictly speaking LES might be initialized by mesoscale models outputs that are previously initialized with GCMs. However, running mesoscale and LES models in cascade presents some different level of complexity in terms of parameterization of variables describing the atmospheric flow evolving at sub-grid size. From a practical point of view we are, at present, far from an operational application level of LES forecasts. More consistent progress in meso-scale forecastings should be achieved to investigate LES forecast performances.

atmospheric turbulence is therefore associated to some dynamic and kinematic characteristics. The optical turbulence is, on the contrary, a concept expressing the turbulence effects on a wavefront (we are therefore in the domain of light/matter interaction). These effects reveal themselves as fluctuation of a passive tracer: the potential temperature θ. The $\Delta\theta$ fluctuations induce fluctuations of the refractive index Δn. Physically speaking, the conditions triggering and developing the dynamic and optical turbulence are not necessarily the same. The most accredited geophysical model for the optical turbulence (Coulman et al., 1995) states that to trigger and develop optical turbulence in an atmospheric vertical slab, two conditions need to be verified simultaneously: (1) $\partial\vec{V}/\partial z > 0$ and (2) $\partial\bar{\theta}/\partial z > 0$. We need therefore a kinematic and a thermal gradient at the same time to trigger optical turbulence. This means that the presence of dynamic turbulence is not necessarily associated to the presence of some optical turbulence. However, the presence of optical turbulence implies the presence of dynamic turbulence.

The Meso-Nh model (Lafore et al., 1998) is based on a system of equations assuming the anelastic approximation permitting us to filter the acoustic waves and correctly represent the gravity waves. The status of the atmosphere is defined with respect to a reference state (hydrostatic equilibrium) with a weak variable potential temperature with respect to the vertical direction. Meso-Nh can therefore simulate all the classical meteorologic parameters ($\vec{V}$, p, T, H,...). The optical turbulence was introduced in Meso-Nh later (Masciadri et al., 1999a) permitting the prediction of 3D maps of the optical turbulence C_N^2 that depends on all meteorological parameters: $C_N^2 = C_N^2 (\vec{V}, T, p, H, L_0)$ where L_0 is the dynamic outer scale. Almost all the other astroclimatic parameters (the seeing ε, the isoplanatic angle θ_0, the wavefront coherence time τ_0, the scintillation rate σ_I^2, the spatial coherence outer scale $\mathcal{L}_0$) depend on the integral of the C_N^2 on the whole troposphere (~20 km) with different weights F as:

$$\int_0^{+\infty} F(h^a, V^b, L_0^c) \cdot C_N^2(h)\, dh \qquad (8.3)$$

(with a, b, and c constant) and they have been integrated in the Meso-Nh model as well (Masciadri et al., 1999a); we refer the reader to that paper for the detailed analytical definition of these quantities (see Preface to this book). Figures 8.3 and 8.4 are example outputs of a typical Meso-Nh simulation done above the site of San Pedro Martir (Baja California, Mexico). Figure 8.3 is a vertical section of the C_N^2 extended over 32 km (North/South). A strong turbulent layer is visible near the ground just above the steep mountain slope on the north side. Such an asymmetry indicates unequivocally the importance of the three-dimensionality of the C_N^2 maps for practical applications. Effects induced by the turbulence on the wavefront can be different if we look at different directions. Horizontal sections (Fig. 8.4) of the same simulation provide us auto-consistent and coherent values for a good astronomical site (a seeing ε~0.6-0.8 arcsec, an isoplanic angle θ_0~1-

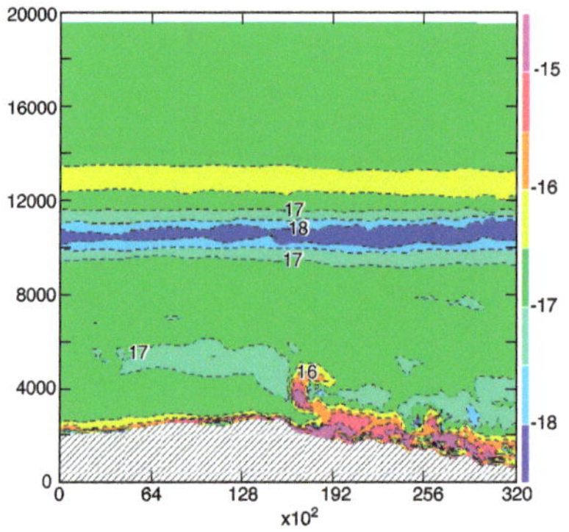

Figure 8.3 Meso-Nh output (ΔT = 3 hours). Vertical section (North/South) of a 3D C_N^2 map in log scale extended on 32 km centered on the San Pedro Martir Observatory (Baja California, Mexico). Blue indicates weak turbulence and pink strong turbulence. Strong turbulence produced above the steep slope of the mountain is visible on the North side.

1.5 arcsec, a wavefront coherence time $\tau_0 \sim$3-3.5 msec, and a spatial coherence outer scale $\mathcal{L}_0 \sim$3-4 m). We highlight that the order of magnitude of the spatial coherence outer scale matches perfectly with recent measurements (Abahamid et al. 2004).

8.3.1 Hydrostatic versus Non-hydrostatic Mesoscale Models

So far we insisted in claiming that non-hydrostatic mesoscale models represent the future of the numerical prediction of the optical turbulence in astronomical applications. The reader might ask why the assumption of non-hydrostaticity is important. The reason is simple. The GCM, models at synoptic scale, usually employ the hydrostatic assumption. This assumes that the pressure in each point is equal to the weight of an air column.

$$\partial p / \partial z = -\rho g \qquad (8.4)$$

This approximation allows us to greatly simplify the equations describing the atmospheric flow evolution as well as the physics of the phenomena we intend to study. Equation 8.4 establishes a relationship between the pressure and the vertical spatial coordinate z that corresponds to an assumption of incompressibility. This means that we conserve the volume and the density. From a physical point, the hypothesis of incompressibility means that the acoustic waves—frequently sources of numerical instabilities—are filtered out. To know when the hydrostatic hypothesis can be applied or not we can refer to the Scale Analysis. The answer fundamentally depends on the model resolution. It has been observed that, for spatial scales smaller than 10 km, the hydrostatic models show tendency to deform the spectrum of the

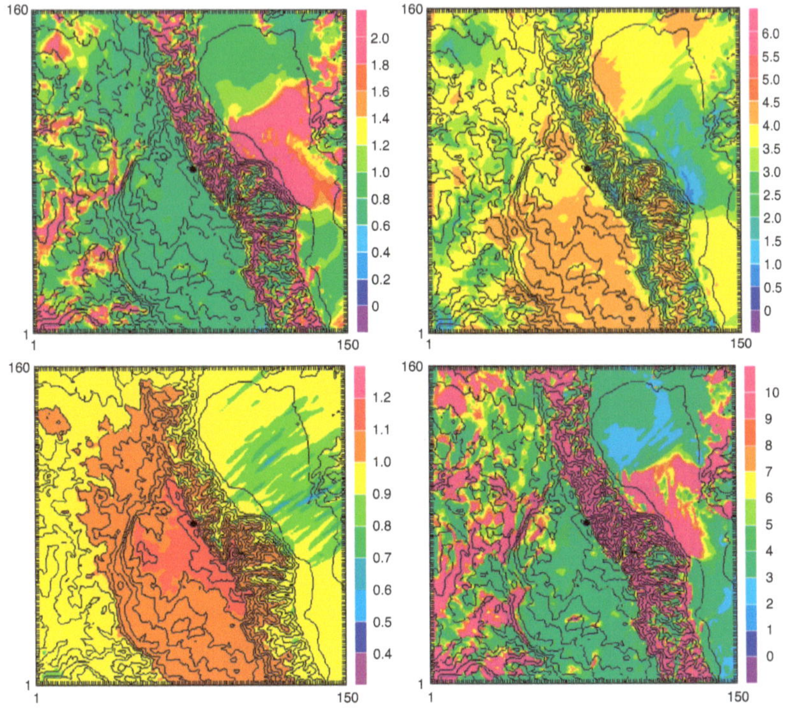

Figure 8.4 Same simulation as Fig. 8.3. Horizontal maps extended 60 km x 60 km with a horizontal resolution of 400 m. Clockwise from the top: the seeing (arcsec), the wavefront coherence time (msec), the spatial coherence outer scale (m), the isoplanatic angle (arcsec). The black lines are the ground levels of the orographic map. The high density of black lines indicates the steepest ground slopes.

gravity waves and, as a consequence, we risk badly reconstructing the orographic waves that are one of the main sources triggering the optical turbulence. This is the reason for choosing non-hydrostatic atmospheric models to reconstruct the optical turbulence.

8.4 Dynamic Turbulence

The turbulence scheme of Meso-Nh (Bougeault and Lacarrere, 1989; Cuxart et al., 1995) is conceived to adapt itself to different spatial scales to allow the model to work at different spatial resolutions. The turbulence scheme is inspired by that proposed by Redelsperger and Sommeria (1981) conceived to have just one free parameter, the mixing length, i.e., the spatial scale on which the speed of a microscopic particle is dispersed. Different mixing lengths can be used in Meso-Nh depending on the scale at which the turbulence is described that is related to the resolution of the system. The turbulence equation can be derived by calculating first, the law of conservation of momentum and expressing each variable $\tilde{X}$ as an average value X plus a random x' with $<x'> = 0$.

$$\tilde{X} = X + x' \tag{8.5}$$

By multiplying the conservation of momentum law by the wind speed of the atmospheric flow we obtain the equation of conservation of the energy. We can divide this equation into two parts: the first describing the temporal evolution of the energy of the averaged flux, the second, the temporal evolution of the turbulence flux. The latter is the equation of the turbulence kinetic energy that in Meso-Nh assumes the form:

$$\frac{\partial e}{\partial t} = -w'u'\frac{\partial U}{\partial z} - w'u'\frac{\partial V}{\partial z} - \frac{1}{\rho}\frac{\partial(\rho w'e')}{\partial z} +$$
$$\frac{g}{\theta}w'\theta'_v - \varepsilon \tag{8.6}$$

where the first and second terms on the right hand side are the shear production, the third term the diffusion and the fifth term the dissipation. The fourth term is the buoyancy term, w' the vertical wind fluctuation, and θ'_v the virtual potential temperature fluctuations.

The turbulence parameterization is done according to the coefficient of turbulence exchange method that allows us to represent the flux of a microscopic quantity ξ as a function of the gradient of the same quantity averaged on a larger spatial size:

$$w'\xi' = K_\xi \frac{\partial \overline{\xi}}{\partial z}, \tag{8.7}$$

and K_ξ is the exchange coefficient, a constant expressing the properties of the turbulent flux in the whole turbulent layer. In Meso-Nh the exchange coefficients are:

$$\overline{w'e'} = -K_e \frac{\partial \overline{e}}{\partial z} \tag{8.8}$$

$$\overline{w'u'} = -K_m \frac{\partial \overline{U}}{\partial z} \tag{8.9}$$

$$\overline{w'v'} = -K_m \frac{\partial \overline{V}}{\partial z} \tag{8.10}$$

$$\overline{w'\theta'} = -K_h\frac{\partial\overline{\theta}_v}{\partial z}. \tag{8.11}$$

The K_h (that we will call simply K) is the key exchange coefficient that will play a role in the calculation of the optical turbulence:

$$K(x, y, z, t) = 0.16L(x, y, z, t)\sqrt{e(x, y, z, t)}\,\phi_3(x, y, z, t) \tag{8.12}$$

where L is the mixing length, e is kinetic energy, and $\phi_3(x, y, z, t)$ is the inverse of the Prandtl number. Considering typical domains for optical turbulence simulations (500 m × 500 m), a 1D turbulence scheme, in which only the vertical turbulent fluxes are considered, is preferable to a 3D turbulence scheme. The mixing length L is defined as the spatial scale at which each particle placed at the level z with a kinetic energy $e(z)$ can move upwards (1_{up}) and downwards (1_{down}) before being stopped by the buoyancy forces. These forces are defined as:

$$\int_{z}^{z+l_{up}} g/\theta_v\,(\theta_v(z) - \theta_v(z'))\,dz' = e(z) \tag{8.13}$$

$$\int_{z-l_{down}}^{z} g/\theta_v\,(\theta_v(z') - \theta_v(z))\,dz' = e(z) \tag{8.14}$$

and L:

$$L = (l_{up}l_{down})^{1/2}. \tag{8.15}$$

The ϕ_3 is a dimensionless function. It characterizes the thermal and dynamic stability of the atmosphere and allows a parameterization of the turbulence at different scales of motion. In 1D dry conditions, $\phi_3(x, y, z, t)$ assumes the analytical form:

$$\phi_3(x, y, z, t) = \frac{1}{1 + C_1 \times (L^2/e) \times (g/\theta_v) \times (\partial\theta_v/\partial z)} \tag{8.16}$$

where $C_1 = 0.139$. The L^2/e behaves in a different way in stable and unstable layers. In very stable layers the mixing length is nearly equivalent to the Deardoff length:

$$L = \sqrt{\frac{2e}{(g/\theta_v) \times (\partial\theta_v/\partial_z)}} \tag{8.17}$$

Replacing Eq. 8.12 in Eq. 8.11 we obtain:

$$\overline{w'\theta'}_\nu = -0.16L \sqrt{e}\,\phi_3 \frac{\partial \overline{\theta}}{\partial z}_\nu \qquad (8.18)$$

8.5 Optical Turbulence Parameterization

Meso-Nh has been adapted (Masciadri et al., 1999a) to simulate the optical turbulence (and its integrated derivatives) according to the relation introduced by Tatarski (1961):

$$C_T^2 = 1.6\varepsilon_\theta\varepsilon^{-1/3} \qquad (8.19)$$

where ε_θ is the rate of temperature variance destruction by viscous process and ε is the rate of energy dissipation related to the turbulence characteristic length L and the energy e by the Kolmogorov law:

$$\varepsilon = C_\varepsilon\, e^{3/2}/L \qquad (8.20)$$

where $C_\varepsilon = 0.7$. The prognostic equation of the variance of the potential temperature $\overline{\theta'^2}$ in the turbulent energy budget is:

$$\frac{\partial \overline{\theta'^2}}{\partial t} = \frac{\overline{\partial w\theta'^2}}{\partial z} - 2\overline{w'\theta'}\frac{\partial \overline{\theta}}{\partial z} - \varepsilon_\theta - \varepsilon_R \qquad (8.21)$$

where ε_θ is the molecular dissipation and ε_R is the radiative dissipation. Assuming that we can neglect the contributions from the triple correlations $w\theta'^2$ and the radiative dissipation we have:

$$\frac{\partial \overline{\theta'^2}}{\partial t} = -2\overline{w'\theta'}\frac{\partial \overline{\theta}}{\partial z} - \varepsilon_\theta \qquad (8.22)$$

The steady state balance equation for the rate of destruction of the variance leads to:

$$\varepsilon_\theta = -2\overline{w'\theta'}\frac{\partial \overline{\theta}}{\partial z} \qquad (8.23)$$

If we replace Eq. 8.23 and Eq. 8.20 into Eq. 8.19 using Eq. 8.18 we obtain C_T^2 expressed as a function of macroscopic variables:

$$C_T^2 = 0.59\phi_3 L^{4/3} (\partial\bar\theta/\partial z)^2. \tag{8.24}$$

Finally, the constant of the structure function of the refractive index is obtained by using Gladstone's law:

$$C_N^2 = \left(\frac{80 \cdot 10^{-6} P}{T^2}\right)^2 C_T^2. \tag{8.25}$$

8.6 Review

We draw an overview of the most important results achieved so far. We refer readers to the references cited for more detailed analysis.

8.6.1 The First Attempts

For the first time in 1986, Chris Coulman urged collaborations between astronomers and meteorologists to solve the problem of the optical turbulence forecast. It was only in 1995 that this invitation was accepted (Bougeault et al., 1995). The authors of the latter paper proposed to use a hydrostatic mesoscale model (PERIDOT) developed in the Centre National des Rechèrches Météorologiques (CNRM/Météo France, Toulouse) coupled with an orographic model with a horizontal resolution of 3 and 10 km to resolve orographic waves produced by the friction of the atmospheric flow over the ground. Results obtained showed some progress with respect to the first attempts of Coulman but they revealed a weak temporal variability of the seeing (ε) and an underestimate of the turbulence due to the relatively low model horizontal resolution. In the same year a different approach was proposed by De Young and Charles (1995). Instead of using a mesoscale atmospheric model the authors used a Direct Numerical Simulation (DNS). This model had a very high resolution (about 10 m) extended over a surface of a few hundreds of meters. The authors tested their model over Mauna Kea (Hawaii) and Cerro Pachon (Chile) sites. The limit of this work was an idealized initialization of the model and a simplified turbulent equation that did not consider the buoyancy forces which are the principal cause of the production of gravity waves. On the other hand it is worthwhile to say that their intention was not to predict the optical turbulence but to study the atmospheric flow evolution for some ideal cases.

8.6.2 The First Non-hydrostatic Mesoscale Model

In 1999, for the first time, it was proposed (Masciadri et al., 1999a,b) to use a non-hydrostatic mesoscale model (Meso-Nh) linked with an orographic model having a horizontal resolution higher than 1 km (typically 400-500 m). For the first time the validation of an atmospherical model was done with measured and simulated C_N^2 profiles and not only integrated values, i.e., the seeing (ε). The study has been applied to Cerro Paranal (Chile), site of the Very Large Telescope

(VLT). The parameterization of the dynamic and optical turbulence in Meso-Nh presented some substantial differences from PERIDOT including the introduction of the function $\phi_3(x, y, z, t)$ (Redelsperger and Sommeria, 1981), an inverse of the Prandtl number taking into account the thermo-dynamic stability properties of the atmospheric flow.

The authors proved that the parameterization of the C_N^2 was possible, i.e., they could reconstruct reliable C_N^2 profiles. This result was far from being obvious because the optical turbulence has typical spatial and temporal scales of fluctuation smaller than those of the standard meteorological parameters and cannot be explicitly resolved by the model over the whole troposphere. For the first time it has been proved that measured and simulated C_N^2 profiles match well from a qualitative (comparable shape) and quantitative (comparable integral over the 25 km) points of view. Masciadri (1998) achieved, among others, two fundamental results:

(1) The horizontal resolution of 1 km is not sufficient to reconstruct reliable vertical distribution of the optical turbulence measured with optical instruments, but a resolution of 400-500 m is sufficient to achieve such a goal. This is true at least for typical mid-latitude sites on the summit of steep mountains.

(2) Internal gravity waves resolved by a non-hydrostatic model several kilometers far away from an astronomical site can propagate along oblique directions affecting the optical turbulence value up to 12 km above the summit (Masciadri et al., 1999a, Fig. 2). This means that thermodynamic instabilities taking place at more than 10 kilometers away from the site can propagate and affect the turbulence conditions up to 10-12 km above the dome of the telescope.

Masciadri (1998) also presented evidence of some limitation of mesoscale simulations, mainly due to initialization data that, in remote regions, can sometimes badly represent the initial conditions. She also discussed possible improvements in the model as the basis for a new calibration method proposed later on (see Section 8.6.3).At the Roque de Los Muchachos Observatory (Canaries Islands) the authors developed a detailed analysis of the ability of the Meso-Nh model in reconstructing nightly evolution of temperature near the surface (Masciadri et al., 2001). It has been proved that simulated temperature well matches the measured temperature (within 3 K) and guarantees a good reconstruction of the energy transfer balance at the interface ground/atmosphere.

More importantly, the authors quantitatively proved in this paper that the particular parameterization implemented in Meso-Nh permits a better estimate of the C_N^2 at the interfaces of stable-unstable regions, which are the regions in which one can expect, with great probability, that turbulence develops. In other words, the Meso-Nh parameterization (Eq. 8.24) is a better parameterization than the classical one commonly used in astronomical context and proposed by Tatarski (1961).

Slightly later, also in the U.S., the importance of the prediction of optical turbulence using atmospheric models above astronomical sites was recognized and the Gemini Observatory initiated the creation of the Mauna Kea Weather Center (Businger et al., 2001; see also Chapter 7).

8.6.3 Calibration, Validation, Reliability

A fundamental step towards the definitive employment of mesoscale models for optical turbulence simulation and validation of the numerical technique has been achieved through a new method of calibrating the Meso-Nh model proposed by Masciadri and Jabouille (2001). This method eliminates some systematic errors and improves the reliability of the model. Although this method has been applied to a specific model (Meso- Nh), the principle on which it is based is universal because it aims to overcome limitations common to all mesoscale models. In that paper the authors presented and discussed the new method and proved, using a small statistical sample of nights, that significant improvements in the reconstruction of the C_N^2 profiles could be achieved. To definitely validate the method, it was necessary to test it for several nights, i.e., on a richer statistical sample of measurements taken in a dedicated site testing campaign. Masciadri et al. (2004) achieved this goal and the Meso-Nh model, calibrated with the method cited, has been statistically validated by comparing simulated and measured C_N^2 profiles obtained during the site testing campaign SPM2000 above San Pedro Mártir (Mexico). This is, at present, a unique site testing campaign conceived to validate the numerical method. To achieve this goal a huge number of different instruments (GS, DIMM, masts, balloons) ran simultaneously for a couple of weeks. The sample used in that paper accurately included only measurements done within a 10 degree cone around the zenith in order to eliminate potential effects due to a lack of spatial homogeneity of the turbulence in a volume. Such a precaution was taken because the goal of this work was precisely the model validation. Figure 8.5 (extracted from Masciadri et al., 2004) shows the averaged C_N^2 vertical profiles simulated by the Meso-Nh model and measured by two different instruments: a Generalized Scidar (GS) and balloons. These are mean values obtained over 10 nights. The measured and simulated C_N^2 profiles of each night are obtained by averaging the C_N^2 profiles over the whole night. The C_N^2 profile measured with the GS for one night is the average of a temporal series of C_N^2 profiles taken at different times during the nights. The simulated C_N^2 profiles are calculated by the model each 2.5 sec (model time-step) and an output is retrieved each 2-minutes. The simulated C_N^2 profile of a single night is the mean of the simulated C_N^2 profiles with a 2 minute time sampling. Table 8.1 reports the seeing in the free atmosphere and the boundary layer calculated for balloons, the Generalized Scidar and the Meso-Nh model. To correctly compare the measured and simulated C_N^2 profiles, the seeing measured by the GS minus the GS dome contribution was compared to the seeing simulated by the Meso-Nh through the whole model atmosphere minus the contribution simulated by the surface (below the telescope height).

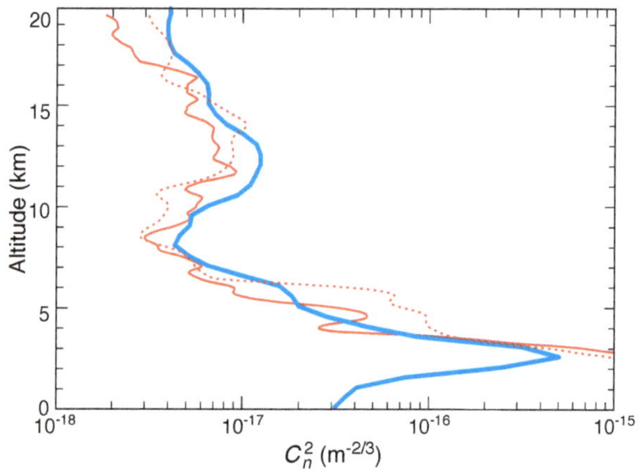

Figure 8.5 Mean vertical C_N^2 profiles measured and simulated over the whole SPM2000 campaign. Solid blue line: Generalized Scidar. Solid red line: radiosoundings. Dotted red line: Meso-Nh model. The relative total seeing obtained from the integration of the C_N^2 profiles is: 0.79 arcsec (GS), 1.07 arcsec (balloons), 0.93 arcsec (Meso-Nh model).

Table 8.1 Summary of the seeing measured and simulated in different regions of the atmosphere during the whole SPM2000 campaign. In the first column is shown the seeing measured by the GS, in the second column the seeing measured by the GS without the contribution provided by the dome, in the third column the seeing simulated by the Meso-Nh model, in the fourth column the seeing simulated by the Meso-Nh model without the contribution provided by the surface, in the fifth column the seeing measured by the balloons.

GS	GS-dome	MNH	MNH-Surf.	Balloons	
0.94	0.62	0.79	0.77	1.00	ε_{BL}
0.42	0.42	0.45	0.45	0.29	ε_{FA}
1.08	0.79	0.97	0.93	1.07	ε_{TOT}

In Masciadri et al. (2004) a detailed statistical analysis is performed by calculating the dispersion between measurements and simulations on single nights and on the whole campaign. The different methods provide, as expected, slightly different percentages in the dispersion between measurements and simulations (~30% in the first case and ~20% in the second). We refer the reader to that paper for a detailed analysis. The main result obtained in that study is that the dispersion between simulations and measurements ($\Delta\varepsilon_{GS,model}$) (GS stands for Generalized Scidar) is comparable to the dispersion between measurements obtained by different instruments ($\Delta\varepsilon_{GS,balloons}$) and, depending on how the statistic is calculated, it ranges between 20% and 30%:

$$\Delta\varepsilon_{GS,model} = \Delta\varepsilon_{GS,balloons} \sim [20-30]\% \qquad (8.26)$$

This result is the best ever reached in terms of model statistical reliability and accuracy of measurements. It is worth highlighting that, in successive studies, a similar dispersion of 30% was measured (Azouit and Vernin, 2005) between a GS and balloons above a different site (Cerro Pachon, Chile) on a sample of 4 weeks. A similar analysis has been done above Mauna Kea (Tokovinin et al., 2005) comparing C_N^2 obtained with two different optical instruments (GS and MASS). If we do not consider the first hundred meters (a comparison would not be fair because the MASS is not or only weakly sensitive[9] to the turbulence) and we look at measurements obtained above 1 km by the two instruments, we can observe that the dispersion between MASS and GS is of the order of 20% above 8 km and of the order of 40-60% in the 1-8 km range. The fact that similar results have been found independently from the site, the instrument (optical and "in situ") and the richness of the statistical sample confirm that the level of accuracy that we can reasonably achieve with measurements is not better than that obtained between measurements and simulations. Masciadri et al. (2004) therefore achieved a benchmark step in the researches of the optical turbulence prediction done with mesoscale models because this result showed that the numerical technique could be considered a reliable technique to estimate C_N^2 profiles and, after a calibration, it can be used, in an autonomous way, to simulate C_N^2 profiles on long time scales. The strength of the result summarized in Eq. 8.26 is that, at present time, we can state that the reconstruction of the optical turbulence made with an atmospherical model such as Meso-Nh is done with an accuracy (dispersion from the true) that is better than that obtained by simulations.[10] This places mesoscale models as a reliable tool for the reconstruction of climatology of the optical turbulence.

For the inexperienced reader it is worth noting that we are talking about statistical comparison, i.e., comparison of median values (simulated and measured) calculated on a single night and/or a set of nights. What we are comparing are typical measured and simulated C_N^2 profiles for each night. No comparison between measurements and simulations is possible, at least at present, on time scale of minutes.

8.6.4 Seasonal Variation Studies with Mesoscale Models

Results obtained in Masciadri et al. (2004) enabled the first seasonal variation study (Masciadri and Egner, 2006) of the vertical distribution of the optical turbulence (C_N^2 profiles) and all integrated astroclimatic parameters (the seeing ε, the isoplanatic angle θ_0, the wavefront coherence time τ_0, the isoplanatic angle

---

[9]The MASS is characterized by a triangle weighting function that decreases the sensitivity of the MASS in a monotonic way down below 500 m up to ground level in which the MASS is not sensitive at all.

[10]Note that the precision of seeing measurements is of the order of 0.01 arcsec, however, the accuracy of measurements, which can be quantified by comparing measurements obtained with different types of instruments, is quantitatively much larger (easily at least by a factor of 10). These two concepts will be discussed in the next section.

for the Multiconjugated Adaptive Optics θ_M where M is the number of deformable mirrors) above the San Pedro Mártir site (Baja California). The authors simulated C_N^2 and wind speed vertical profiles for 80 nights uniformly distributed in one year (2001). From the C_N^2 and the wind speed vertical profiles they calculated all the integrated astroclimatic parameters commonly used to characterize an astronomical site and included the optimal altitude at which to conjugate the Deformable Mirrors (DMs) of a MCAO system having 1, 2 and 3 DMs. That paper provided new insights on the nature of the optical turbulence and it showed the potential of the non-hydrostatic atmospheric models for optical turbulence simulations. In one shot, a 3D simulation can draw a complete characterization of all parameters useful to characterize the turbulence above an astronomical site.

For the first time the seasonal trend of the C_N^2 at different heights was studied (Fig. 8.6). In that figure, the typical C_N^2 peak in the free atmosphere becomes weaker and shifts at higher heights in the summer time (north hemisphere). The authors could not compare the morphology of this particular feature in the simulated and measured C_N^2 profiles above San Padro Mártir because they had no seasonal coverage of measured C_N^2 profiles above this site. However, a similar trend in the measured C_N^2 profiles has been recently observed (Fig. 8.7) above Mt. Graham (Arizona) (Egner et al., 2007), an astronomical site placed not far away from San Pedro Mártir (Baja California). There are, therefore, realistic possibilities that such a trend can be typical of (many) other sites. Masciadri and Egner (2006; Section 3.2) provided a physical explanation for such a seasonal trend. Briefly, they proved that the probability to trigger thermodynamic instabilities is lower in summer than in winter time at these heights in that site. An increase of the median wind speed shear of ~ 17-20 km (as observed by Masciadri and Egner, 2006) can justify the shift of the C_N^2 peak at higher heights with respect to the winter time. It would be interesting to verify if similar trends are observed in other sites and if this explanation can be extended to other sites. Such an effect will probably be

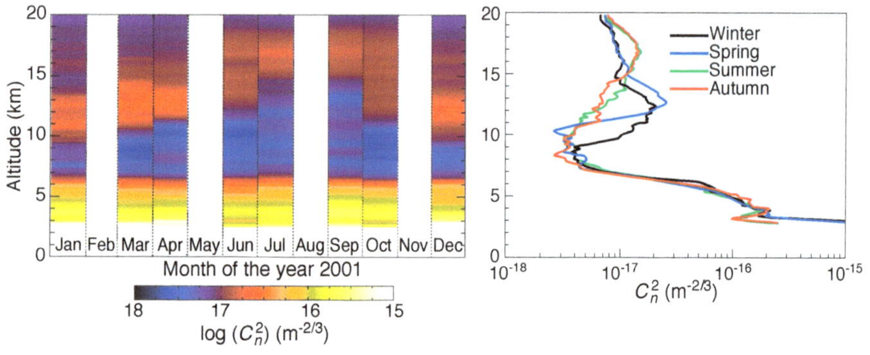

Figure 8.6 Left: C_N^2 seasonal evolution versus 80 nights uniformly distributed in the year 2001. Right: Median vertical C_N^2 profiles in winter, spring, summer, and autumn.

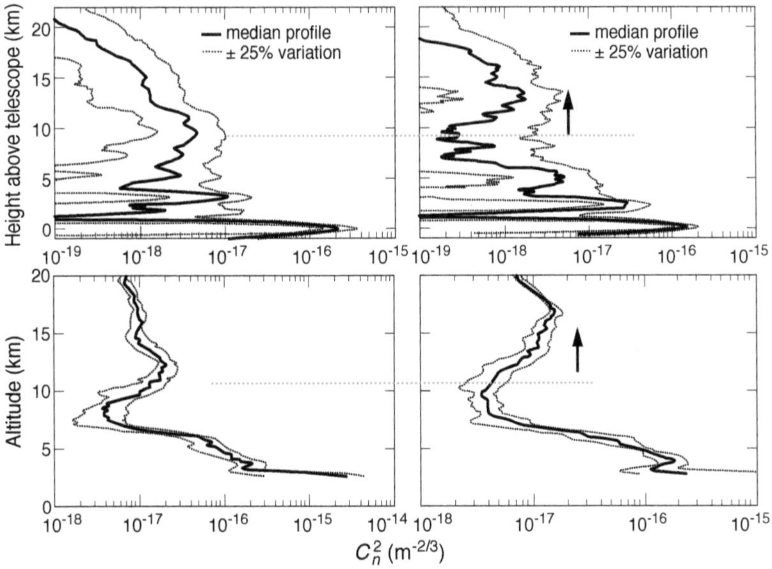

Figure 8.7 Top: Median C_N^2 profiles measured above Mt. Graham in winter (left) and in summer (right). From Egner and Masciadri (2007). Bottom: Median C_N^2 profiles simulated above San Pedro Mártir in winter (left) and in summer (right). From Masciadri and Egner (2006). A shift of the C_N^2 peak in the free atmosphere towards greater heights can be observed above both sites during the summer time.

barely visible with a MASS/DIMM due to the relatively low vertical resolution at those heights. A GS is certainly a more sensitive instrument to resolve such an effect.

Concerning the seasonal variation of integrated parameters, Masciadri and Egner (2006) proved that the seeing seasonal variation is mainly affected by the boundary layer seasonal variation and the isoplanatic angle θ_0 increases in summer time as a consequence of the weaker C_N^2 in the free atmosphere in this season. A further encouraging result is that simulations, done on such a large number of nights (80), provide a median seeing (0.90±0.01 arcsec) in good agreement with measurements done above the sites with different instruments (0.87 arcsec)[11] (Masciadri and Egner, 2006). The same can be said for the seeing in the 10-15 km slab $\varepsilon_{[10,15],sim}$ = 0.22 arcsec (Masciadri and Egner, 2006) versus the measured $\varepsilon_{[10,15],meas}$ = 0.24 arcsec (Avila et al., 2004). The median simulated isoplanatic angle (θ_0=1.42 arcsec) is well correlated with the measured one (θ_0=1.87 arcsec; Avila et al., 2004). Estimates on the isoplanatic angle matches within ~20-25%. The seasonal variation of the simulated seeing $\Delta\varepsilon_{sum-win}$ ~ 0.22 arcsec (Masciadri

---

[11]If we include in the sample of measurements, the recent results obtained in the context of the survey made by the TMT project at San Pedro Mártir (Schoeck et al., 2008) the median seeing is 0.86 arcsec.

and Egner, 2006) is well correlated with the seeing measured with a DIMM (integrated values) (Michel et al., 2003; $\Delta\varepsilon_{sum-win} \sim 0.22$ arcsec) and the seeing measured with a SST (integrated values) (Echevarria et al., 1998; $\Delta\varepsilon_{sum-win} \sim 0.11$ arcsec). The atmospheric circulation in this vertical slab is strongly affected by the jet-stream and, considering that the synoptic circulation has some seasonal trends, results obtained indicate that such seasonal features are reflected also on a physical phenomenon characterized by smaller spatial and temporal fluctuation scales such as the atmospheric turbulence.

A further interesting conclusion in Masciadri and Egner (2006) is related to the quantification of the parameters characterizing the Adaptive Optics applications such as the optimal heights (H_M) at which to conjugate the deformable mirrors for the Multi Conjugated Adaptive Optics (MCAO) and the isoplanatic angle for the MCAO (θ_M). We cite here just one among the outputs of this study. The isoplanatic angle obtained tuning the optimal conjugated heights of the DMs each nights is, in general, larger than the isoplanatic angle obtained with a fixed conjugated height. The authors proved that by taking different optimal heights in summer and winter it is possible to maintain this loss within 20% with respect to keeping a fixed optical height.

8.6.5 First Evidence of the Finite Horizontal Extent of Turbulence Layers

The possibility to reconstruct 3D C_N^2 maps offers to the numerical technique some incredibly useful intrinsic potentialities. In this section we will deal about one of these.

It is well known that the optical turbulence is vertically distributed in thin layers of a few tens of meters (Azouit and Vernin, 2005) but not much was known about the horizontal size of turbulent layers, at least up to recent years. The theory (Roddier, 1981) applies the approximation of infinite horizontal layers and the same assumption is done for the data reduction of measurements retrieved from all the optical instruments measuring integrated turbulence values or its vertical distribution. These assumptions are justified by the fact that horizontal fluctuations are smaller than vertical fluctuations.

However, some studies (Masciadri, 2000; Masciadri et al., 2001) showed for the first time that, integrating the C_N^2 along lines of site different from zenith, the integrated seeing ε can be substantially different. Figure 8.8-bottom shows a horizontal map of the seeing ε simulated above the San Pedro Mártir site (Baja California) (Masciadri, 2000) on a surface of 60 km x 60 km and a horizontal resolution of 400 m. Each point of this map represents the seeing obtained by integrating the volumetric C_N^2 along lines of sight parallel to the zenith. This is a very practical system because it permits a comparison of simulations versus measurements. Measurements provided by optical instruments are indeed obtained assuming that the horizontal layers size is infinite with homogeneous morphology.

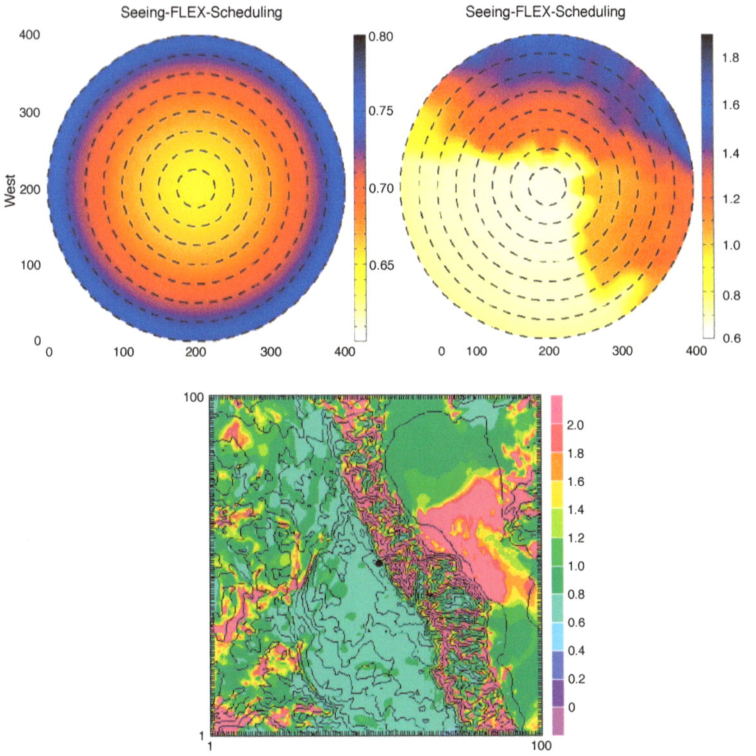

Figure 8.8 Bottom: Horizontal map of the seeing simulated above the San Pedro Mártir site (Baja California). The extension of the surface is 60 km and the horizontal resolution is 400 m. Top-left: Seeing polar map obtained integrating the C_N^2 above the site (central black dot) along the zenith. Each circle shows the seeing computed integrating at a solid angle different from 0, having an increment of 5 degrees and a maximum angle of 40 degrees. The C_N^2 is horizontally uniform. Top-right: Same as left side but using the 3D C_N^2 simulated by the model and used to retrieved the seeing map in Fig.8 bottom.

The seeing retrieved from the instruments are corrected by the air mass factor that consists of replacing the coordinate z with ($z \cdot \sec \gamma$) where γ is the angle between the zenith and the line of sight.

But what would happen if we integrate the C_N^2 along lines of sight different from zenith? To put in evidence the consequences of such a calculation and to quantify the potential differences obtained by accepting or rejecting the assumption of horizontal homogeneity of the turbulence layers, Masciadri et al. (2002) proceeded in the following way. Figure 8.8-top-left shows what the authors call a seeing polar map. The center of the map is the seeing obtained integrating the C_N^2 along the zenith above the San Pedro Mártir site (black dot in Fig. 8.8-bottom). Each circle shows the seeing computed by integrating the C_N^2 at a solid angle different from 0, having an increment of 5 degrees and a maximum angle of 40 degrees. The C_N^2 horizontal distribution is taken uniform and equal to the C_N^2 on the vertical direction above the Observatory i.e., the black dot. The seeing polar map has, therefore, the shape of a parabola having the minimum over the site. This reproduces the procedure used to retrieve measurements with optical instruments. Figure 8.8-top-right shows the same seeing polar map but obtained by integrating the 3D C_N^2 distribution simulated by the model and producing the seeing map shown in Fig. 8.8-bottom.

Two important results can be retrieved by these pictures:

(1) The assumption of a uniform C_N^2 is a strong assumption. At least in this case, in the northeast direction, for example, we find difference in seeing with respect to the case represented in Fig.8-top-left that can attain 1 arcsec.

(2) If we look at two different directions above an astronomical observatory the seeing ε can attain difference that can reach 1 arcsec.

If we think about the consequences that such a fact can have on the management of a telescope we conclude that 3D C_N^2 maps appear necessary to predict the seeing values on a field of view of +/- 60° with respect to the zenith. Figure 8.8 was an interesting and pioneering result but prudence suggested another look for an independent result. In the same paper the authors provided a first evidence of the first finite extent of the horizontal size of turbulent layers observed not only with simulations but also with measurements (Generalized Scidar). In that context they observed a maximum seeing difference of 0.3 arcsec within 40° that is enough to produce some consequences from an observational point of view. Masciadri et al. (2002) discussed an extended list of serious implications that this fact might have for the AO techniques. A detailed study of the horizontal extent of the turbulent layers horizontal size is therefore necessary although challenging. From a practical point of view the next step would be to quantify which is the typical difference in seeing that can be obtained within 40° around the zenith in statistic terms. This would require the organization of a dedicated site testing campaign extended on long time scale.

8.6.6 Surface Layer Wind Speed Prediction with Mesoscale Models

Mesoscale models can predict another key parameter extremely critical for the new generation ground-based astronomy: the near ground wind speed (NGW). Many ELT projects exist at present. Because of the huge size of these facilities, the selection of an optimal site with low wind speed near the ground is a major priority to ensure the stability of the telescope structure. Wind speed estimates provided by General Circulation Models (GCMs) from Meteorological World Centers are hardly reliable because of the low horizontal resolution (~0.5° up to 2.5°) of the GCMs. In Masciadri (2003) a dedicated study was carried out using a mesoscale model (Meso-Nh) with a horizontal resolution of 1 km (~30 arcsec). Two astronomical sites have been considered as a benchmark test: Cerro Paranal

(Chile) and Maidanak (Uzbekistan). Measurements of the near ground speed taken during some years above the two sites show a typical wind speed difference of 4-5 m s^{-1} (Masciadri, 2003) and Maidanak was revealed among the sites with the weakest surface wind speed in the world. In that paper measurements of the NGW for 20-25 nights at each site are compared to simulations obtained from the mesoscale model and a GCM and a statistical analysis is performed. The author calculates the cumulative distribution of the relative error (CDRE) with respect to measurements for the GCM and the Meso-Nh model. The most important results achieved are these:

(1) The GCM definitely fails in discriminating between the NGW above the two sites. More precisely, the mean wind speed reconstructed above Maidanak is higher than above Paranal. The mesoscale model Meso-Nh can discriminate the wind speed above the two sites. It can recognize a weaker wind speed above Maidanak than above Paranal.

(2) The cumulative distribution of the relative error (CDRE) between measurements and simulations calculated with Meso-Nh gives an excellent CDRE = 26.84% for Maidanak and a CDRE = 47.30% for Paranal. The score of success is definitely better than that obtained by GCMs: a very poor CDRE = 99.40% for Maidanak and CDRE = 66.67% for Paranal. Such a high percentage tells us that the GCMs are not at all reliable for the NGW estimates.

Masciadri (2003) proved, therefore, that mesoscale models can discriminate and identify sites having particularly weak wind speed near the ground while GCMs cannot. Moreover, the relative error Δr between the measured and simulated wind speed near the ground obtained by mesoscale models is 27-47% less than that obtained by the General Circulation Models. The latter are only weakly reliable in proximity to the surface because orographic effects are poorly represented by such models.

8.6.7 Recent Studies—Grid-nesting

The grid-nesting technique, which allows a mesoscale model to increase the resolution on small domains while still keeping the atmospherical flow reconstructed at lower resolution on large domains, is more reliable in recent years, allowing more applications of such a typology of models. The grid-nesting configuration consists of a set of imbricated domains. The first coarse grid size domain extended on a large surface is linked to a domain extended on a smaller surface and higher horizontal resolution overlapping a portion of the large domain and so on in a concatenated sequence fixed by the user. This configuration was implemented in studies related to astronomical applications: Cherubini et al. (2008) completed a study above Mauna Kea (Hawaii) employing a non-hydrostatic mesoscale model (MM5[12]) and a grid-nesting configuration (four domains with 27, 9, 3, and 1 km horizontal resolution). A comparison of measurements/simulations has been provided on a few selected number of nights. It has been shown that the model can reconstruct the C^2_N but the relative errors are still relatively large and there is still not a statistical validation of the ability of the model in reconstructing the vertical distribution of the optical turbulence that, as shown in Masciadri et al. (2004), is fundamental for the model reliability. On the other hand, the focus of this study is the fact that many technological aspects related to the operational applications have been considered. The model runs indeed on a supercomputer on the Mauna Kea Observatory and provides nightly predictions. In the same period Masciadri's group, in the context of the ForOT[13] project, started a study of the characterization of the atmosphere and turbulence developed above the Internal Antarctic Plateau (Lascaux et al. 2008, 2010)[14], a continent that has some interesting potentialities for astronomical applications, namely a particular weak turbulence above the first 30-40 m. In this study the Meso-Nh model has been run in a grid-nesting configuration of three domains: 25, 5, and 1 km, and meteorological parameters as well as optical turbulence features have been compared to measurements[15] taken in winter at Dome C. The authors showed that the grid-nesting configuration with a high resolution of 1 km provides better correlation with measurements than the configuration with a horizontal resolution of 100 km as used by Swain and Gallée (2006) with a mesoscale model MAR (Gallée and Schayes, 1994). The Meso-Nh model has been validated and the authors showed that Meso-Nh is able to reconstruct reliable values for the three most important parameters that characterize optical turbulence: (1) the surface layer thickness; (2) the seeing in the free atmosphere; and (3) the seeing in the surface layer. Also they showed that the grid-nesting configuration with the high resolution of 1 km provides better correlations. This means that, even with a relatively flat topography typical of these regions, the grid-nesting configuration is necessary to perform optical turbulence simulations.

8.7 Measurements and/or Simulations

Measurements and simulations are tools that can be considered complementary with respect to the characterization of the optical turbulence. They can be seen as tools that answer different questions.

---

[12]MM5 is developed by NCAR, Boulder Colorado, US

[13]ForOT: http://forot.arcetri.astro.it

[14]We highlight the interesting technical information related to this study: simulations have been run on a 64-bit powerful workstation and not a supercomputer. This certainly opens to this research field new perspectives of development.

[15]Measurements of meteorological parameters refer to the Osservatorio Meteo Climatologico of the Programma Nazionale di Ricerche in Antartide (http://www.climantartide.it) operated by PNRA and IPEV. Optical turbulence measurements are taken from Trinquet et al. (2008).

Simulations:
- provide 3D C_N^2 maps
- provide forecasts of the optical turbulence
- provide complete climatology of the optical turbulence and its integrated derivatives extended on time scale of the order of decades
- are extremely less expensive and faster than measurements

Measurements:
- provide real-time estimates
- access all spatial and temporal scales typical of the turbulence that changes on time scales of the fraction of arcseconds

These concepts need to be taken and interpreted with some caution to avoid erroneous conclusions. One of the most common and dangerous habits in our scientific community has been to consider a measurement a more accurate estimate than a model calculation. It is true that a calculation done with a model is, by definition, a representation of reality, however, a measurement does not represent, by default, the 'true value'. As with all the random parameters, the turbulence has to be (spatially and temporally) quantified in statistical terms. Researchers working in this research field frequently expect/pretend that calculations have levels of accuracy that no measurements could achieve so far. To pursue this discussion, we stress a few fundamental statistical concepts. The concept of accuracy is strictly related to the concepts of veracity and it tells us how close the estimate is to the true value, which in the case of our applications, is not known. The concept of precision is related to the capacity of an instrument to reproduce the same result if the estimate is repeated in the same conditions several times. Precision is independent from agreement with the true value and it expresses the uncertainty of an estimate and ultimately the statistical error $\sigma/\sqrt{N}$. The reliability of an estimate is related to its accuracy. It is relatively easy to achieve high level of precision that scales as $\sigma/\sqrt{N}$ where σ is the standard deviation and N is the independent number of estimates. It is much more difficult to have a high level of accuracy in the case of a random and highly non-linear parameter such as turbulence. Good precision is necessary but it is not sufficient to have a reliable (i.e., accurate) estimate of a parameter. An estimate can be very precise but completely wrong, i.e., not accurate. We are ultimately limited by the systematic errors and the non-statistical errors. The calibration of an instrument as well as of a model is a necessary step to reduce some systematic biases and/or uncertainties due to free parameters and it is necessary to achieve better levels of accuracy.

The concept of calibration has been neglected for many years. Luckily, in recent years, greater attention has been shown toward this aspect for both measurements and simulations. In some cases, bias in measurements has been discovered and studied in a systematic way, and solutions or methods to control the identified problem have been proposed. For example, it has been observed by several authors that an exposure time of 5 or 10 msec provides underestimates of the seeing measured with a DIMM due to filtering of a part of the turbulent spectrum frequencies. This effect has been formalized in Tokovinin (2002). The sensitivity and statistical noise in the Generalized Scidar have been studied by Prieur et al. (2001). A method to overcome the frequent saturation of the scintillation in the MASS has been recently proposed (Tokovinin and Kornilov, 2007). In other cases it has been observed that the DIMM, commonly used to measure the τ_0 from the calculation of r_0 and the wind speed at 200 mb, provide measurements with relative errors of the order of 20-50 % up to 60% (Masciadri and Egner, 2006). All methods aiming to quantify the τ_0 and based on the measurement of integrated parameters suffer such a limitation. On the other hand, Masciadri and Jabouille (2001) have been pioneers in the introduction of the calibration in the mesoscale models for the optical turbulence simulations.

If some improvements have been achieved in the direction of the identification of biases and systematic errors, it still remains mostly difficult/impossible to select a reference as the true value. At present, the usual approach to estimate the true value is to calculate the dispersion of estimates of the same parameter (the seeing ε for example) obtained with different instruments and to define the true within an interval of uncertainty $\Delta\varepsilon$ that mathematically expresses the concept of accuracy. If one refers to results reported in the literature on the estimates of the optical turbulence above an astronomical site it is possible to observe that the typical statistical error is of the order of a few hundredths of arcsec (± 0.01 arcsec). However, it is a fact that, at present, levels of accuracy (for both measurements and simulations) of the integrated astroclimatic parameters have not yet been observed to be better than 20-30%. As mentioned in Section 8.6, Masciadri et al. (2004) found a discrepancy of the order of 20-30% between balloons and a GS. Similar dispersion has been calculated with a richer statistical sample and in a different site with the same instruments by Azouit and Vernin (2005). A comparison done between two optical instruments, a MASS and a GS (Tokovinin et al., 2005), gives relative errors of the order of 20% above 8 km but achieves 40-60% in the 1-8 km range. This simply means that there is still room for refinements in the experimental techniques, in the identification of systematic errors, and in calibration procedures in this research field. This is not surprising considering that turbulence is certainly one of the most difficult parameters in physics to be quantified.

For the C_N^2 simulations obtained with atmospheric models or with vertical profilers (instruments), we can state that the calibrations need to be done for the vertical distribution of the optical turbulence (C_N^2 vertical profiles) and not for integrated values such as the seeing (ε). This is the only way to identify the behavior of the tool of investigation (model or vertical profiler) in the different vertical slabs of the troposphere where turbulence differs: surface layer, boundary layer, and free atmosphere. Calibrations done on integrated values (the seeing

ε) do not necessarily guarantee the correctness of the vertical distribution of the turbulence in the atmosphere.

For the numerical calculations it is extremely important to establish criteria defining the score of success of a prediction. At present, the best method is to compare the difference between measurements and simulations with the discrepancy between measurements provided by different instruments. This criteria allows us, when the dispersion measurement/simulation is not worse than the dispersion between measurements, to use mesoscale models to characterize the optical turbulence above astronomical sites in an autonomous way. This does not mean, of course, that a better accuracy cannot be reached in the future with simulations. It is evident that, according to the progress in our ability to quantify the accuracy of both measurements and simulations, more sophisticated algorithms to evaluate the performance of forecasts should be developed in the future.

8.8 Perspectives

Research carried out in recent decades showed that we can reproduce the optical turbulence vertical distribution in the atmosphere with non-hydrostatic mesoscale models. It has been successfully proved that it is also possible to achieve reliable results in statistical terms and that these models can be used to characterize most of the turbulence features in a reliable way. We saw that, frequently, the models supply complementary information with respect to measurements. In some cases, operational approaches were tested. Which are the critical goals that still need to be achieved in this research field? In our opinion, there are two important challenges. First, we need to prove the ability of mesoscale models to forecast the optical turbulence. Second, we need to prove the ability of mesoscale models in discriminating the quality of the sites. The results obtained so far (Masciadri et al., 2004; Masciadri and Egner, 2006) showed that it is possible to address these challenges. Related target topics include the definition of strategies for the quantification of a score of success for a model running in an operation modality and the practical employment of such a technique. Progress in this discipline will strongly depend on improved data assimilation procedures for General Circulation Models and mesoscale models. The success of this approach will depend on our ability to interface astronomy with meteorology.

To achieve our two main objectives, the best approach would be to carefully plan campaigns of measurements done with several instruments running simultaneously and monitoring different part of the troposphere with a higher level of resolution in proximity to the surface and a lower resolution in the free atmosphere. We should use as many different instruments as possible based on different principles. It is important to understand that the strategy of site testing campaigns planning could change in the future and respond to the necessities of the modeling science. This process is fundamental to create benchmark cases for these studies.

It is also evident that this research field needs to pass to a new dimension to exploit its potential. We need to move toward operational applications, i.e., flexible scheduling of scientific programs and the selection of instruments to be placed at the focus of telescopes. But this depends on both the research and the technology. Astronomers need to use the tools of investigation (the atmospheric model) typical of a different discipline: the physics of the atmosphere. The approach of the ForOT project is to develop further and new competences in astronomy. Its basic aim is to enlarge the field of the High Angular Resolution Techniques and Turbulence Characterization employing measurements and numerical calculations obtained with mesoscale atmospheric models. This means running the models on PC clusters and workstations, not necessarily on super-computers in meteorologic centers. In other words, at a certain point it will be necessary for the research approach to pass to an operational one. From the point of view of computational requirements and available resources, this is now feasible. Figure 8.9 schematically represents the goal of this project at long time scales. The High Angular Resolution (HAR) techniques, which were developed in the field of astrophysics, is adaptive optics-interferometry, wavefront perturbations induced by the atmospheric turbulence, and image

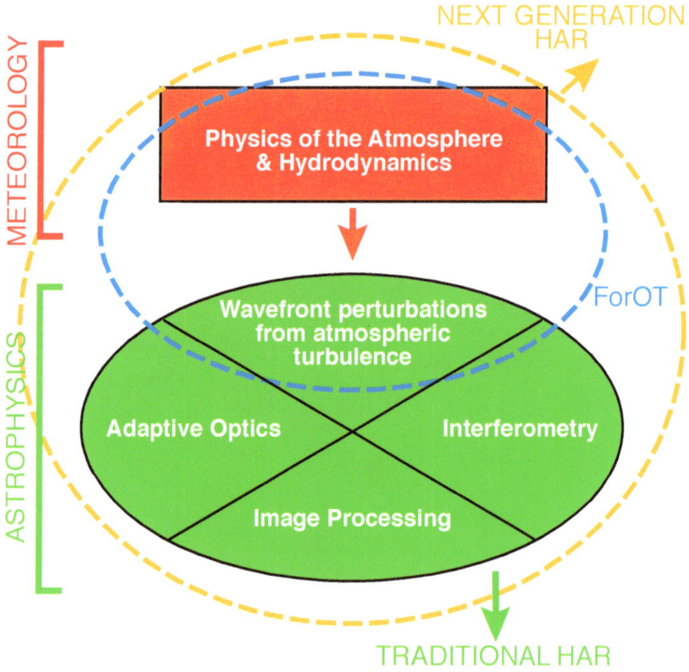

Figure 8.9 Schematic representation of the Manifesto of the ForOT Project. See text in Section 8.8. From http://forot.arcetri.astro.it.

processing. ForOT aims to integrate into the traditional High Angular Resolution (HAR), a research line typical in meteorology: the physics of the atmosphere and the hydrodynamic. The final goal is to establish the basis for a new HAR generation and the formation of a new class of interdisciplinary scientists.

Acknowledgments

This work is funded by a Marie Curie Excellence Grant (FOROT) MEXT-CT-2005-023878-FP6 Program.

Authored by
E. Masciadri
INAF—Osservatorio Astrofisica di Arcetri, Florence, Italy

CHAPTER 9

An Operational Perspective for Modeling Optical Turbulence

Ground-based astronomy depends heavily on the impact of the weather on the observing conditions. While clear weather is the minimum requirement for astronomers to be able to conduct productive optical observations, the optically turbulent state of the atmosphere is also a primary concern, since turbulence greatly impacts the resulting image quality and therefore the range of possible observations. Turbulence causes amplitude and phase fluctuations in electromagnetic waves propagating through the atmosphere, constraining the maximum telescope resolution and resulting in telescope image degradation. Adaptive optics (AO) allows for a partial correction of the image degradation caused by atmospheric turbulence (Beckers, 1993). However, to maximize the benefits of AO, knowledge of the vertical profile of optical turbulence is needed (see also Chapter 6).

This chapter describes the implementation, validation, and calibration of an atmospheric turbulence algorithm applied in an operational mesoscale numerical weather prediction model at the Mauna Kea Weather Center (MKWC). The operational challenges of using the model to fine-tune the forecast of the optical turbulent state of the atmosphere at the astronomical sites at the summit of Mauna Kea are also addressed.

9.1 The Mauna Kea Weather Center Forecast System

The Mauna Kea Weather Center (MKWC) is a joint program between the Department of Meteorology and the Institute for Astronomy at the University of Hawaii (see also Chapter 7). The main task of the MKWC is to provide custom weather support to the community of astronomers who operate the observatories on the summit of Mauna Kea. Since January 1999, the MKWC has issued two daily (10am and 5pm) summit weather forecasts for the Mauna Kea Observatories (Businger et al., 2002). Each forecast extends over a 5-day period, with a less detailed outlook 6-10 days into the future. In constructing the forecasts, the MKWC forecaster uses as guidance the output from global models at the national operational centers, the synoptic and asynoptic observations available in real-time, and the output from the mesoscale model run at the MKWC. The custom forecasts, along with satellite and radar data, model output, and summit weather observations are made available to the astronomers, and the public, via the MKWC web site (http://mkwc.ifa.hawaii.edu). The MKWC forecasts include a range of meteorological variables of importance to the operation of the summit observatories, such as trends in summit temperature, wind speed, cloudiness, and precipitable water (Businger et al., 2002). However, the focus of this chapter will be the implementation of an optical turbulence algorithm and C_N^2 /seeing forecasts.

9.1.1 Overview of Hardware and Data Flow

The MKWC staff maintains a complex hardware and software configuration that handles the data flow, model input/output, and serves products to the web pages (Fig. 9.1). Our primary data source, including model output from the National Centers for Environmental Prediction (NCEP), is Unidata, which is located in Boulder, Colorado, and is funded by the National Science Foundation. Unidata distributes geosciences data sets and software applications to handle and display these data. In addition to the data stream provided by Unidata, special satellite data sets are received directly from the National Oceanic and Atmospheric Administration (NOAA) and the National Weather Service (NWS).

The diverse observational data and model output are combined in the data assimilation system; currently Local Analysis and Prediction System (LAPS) (McGinley 1989) runs on our primary server to produce and provide initial and boundary conditions for our mesoscale model as described in Cherubini et al. (2006). The mesoscale model, which is described in the next section, runs on a dedicated Linux cluster.

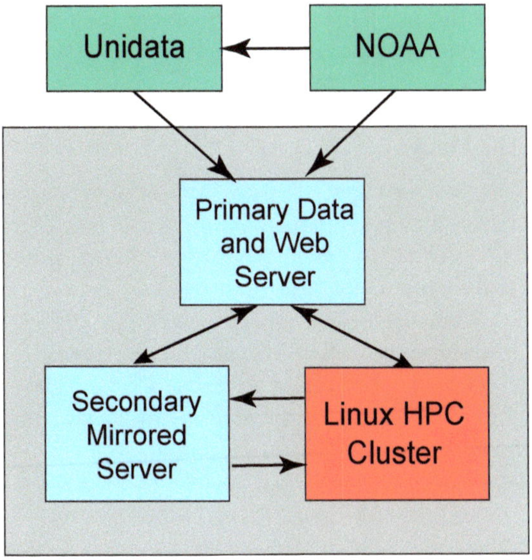

Figure 9.1 Schematic flow chart of the hardware and data flow distribution for the MKWC. Green boxes indicate data sources. Light blue boxes indicate the servers involved in data management, web serving, and pre- and post-processing for WRF. The red box indicates the computer cluster that runs WRF.

To make the operation of the MKWC more robust, our primary server is mirrored with a secondary server that duplicates data management and web serving functionality. When one server has a hardware or software issue, the other server automatically takes over the task of making data available to users via the Internet, thus minimizing disruption in availability of MKWC web products.

9.1.2 Overview of Mesoscale Model

From 1999 to the beginning of 2008, the MKWC implemented the fifth-generation Mesoscale Model (MM5) that was originally developed at Pennsylvania State University and National Center for Atmospheric Research (Grell et al., 1985; Cherubini et al., 2008a, b). Early in 2008, the MKWC staff transitioned from MM5 to the Weather Research and Forecasting (WRF) model (Klemp et al., 2007, http://www.wrf-model.org), the next-generation mesoscale numerical weather prediction system, which had become sufficiently stable and mature for our operational needs.

The WRF is a non-hydrostatic model with multiple nesting capabilities. The current MKWC configuration encompasses 3 two-way nested domains, with horizontal resolution of 15, 3, and 1 km, corresponding to time steps of 75, 2, and 5 seconds respectively (Fig. 9.2). The outermost domain covers a wide portion of the central Pacific area, the 3-km resolution domain spans all of the larger Hawaiian Islands, and the innermost domain covers the entire island of Hawai'i (Big Island). Forty levels in the vertical are used; the vertical spacing is on the order of tens of meters for the levels nearest the ground and gradually increases with height as shown in Fig. 9.2. The model top is fixed at 10 mb, which corresponds to a height of about ~25 km above the ground level.

The WRF physics package used in the current operational configuration includes (i) a 3-class simple ice scheme for the microphysics scheme (Hong et al., 2004), (ii) the Kain-Frietsch cumulus convection scheme (Kain and Fritsch, 1990; Kain and Fritsch, 1993) for the 15- and 3-km domains, (iii) the Mellor-Yamada-Janjic (MYJ) planetary boundary layer scheme (Janjic, 2002), which solves the prognostic equation for the TKE, and (iv) the RRTM longwave-shortwave radiation scheme (Mlawer et al., 1997).

The WRF model is run twice daily with 0000 and 1200 UTC initial conditions produced by LAPS, as discussed in Section 9.1.1. Boundary conditions are updated every six hours using model output from the National Centers for Environmental Prediction Global Forecasting System.

9.2 The Optical Turbulence Algorithm

The prognostic algorithm for atmospheric turbulence previously implemented in the MM5 model calculates the refractive index structure function C_N^2, which describes the turbulent fluctuations of the atmospheric refractive index (Cherubini

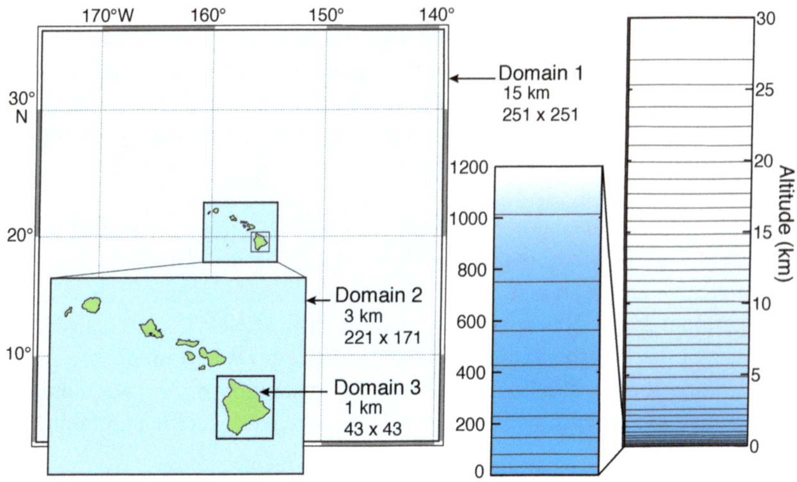

Figure 9.2 Configuration of the nested grids in WRF. Inset map shows an expanded view of the main islands in the Hawaiian chain. The vertical resolution of the model is depicted in the diagram at right, including an expansion of the lowest 1200 meters of the model domain.

et al., 2008a, b). The integral of C_N^2 along the line-of-sight of a telescope is proportionally correlated to a parameter referred to as the telescope *seeing*. Seeing describes the impact of the turbulent fluctuations on image quality at the telescope (see Preface to this book).

The optical turbulence algorithm implemented within WRF differs somewhat from the one implemented in MM5. The boundary layer schemes in WRF and MM5 differ and these schemes provide information on the turbulent kinetic energy (TKE) profiles that is critical for the calculation of the C_N^2. The optical turbulence implemented in WRF is calculated following Masciadri et al. (1999a, b). In particular, the C_N^2 profile is calculated using the following relation (Masciadri et al., 1999):

$$C_n^2(z) = \left(\frac{80 \times 10^{-6} p}{\vartheta^2}\right)^2 C_\vartheta^2(z) \qquad (9.1)$$

where the temperature structure coefficient C_ϑ^2 is calculated according to

$$C_\vartheta^2(z) = a^2 \left(\frac{K_h}{K_M}\right) L_0^{4/3} \left(\frac{\partial \vartheta}{\partial z}\right)^2. \qquad (9.2)$$

In equation (2), K_H and K_M are the exchange coefficients for heat and momentum, a is an empirical constant and L_0 is the outer length scale of turbulence (the upper bound of the inertial sub-range), which, in stable conditions can be calculated as:

$$L_0 = \frac{0.76\sqrt{e}}{N},$$
(9.3)

where N is the Brunt-Vaisala frequency and e is the TKE. For unstable conditions L_0 is calculated as follows:

$$L_0 = (l_{up}l_{down})^{1/2}$$
(9.4)

where l_{up} and l_{down} are calculated by solving:

$$\int_{z}^{z+l_{up}} \frac{g}{\vartheta}(\vartheta(z) - \vartheta(z'))dz' = e(z)$$

$$\int_{z-l_{down}}^{z} \frac{g}{\vartheta}(\vartheta(z') - \vartheta(z))dz' = e(z)$$

The MYJ scheme includes a background TKE of 10^{-1} m^2 s^{-2}. In order for turbulent production to begin under conditions of very stable atmosphere, the turbulent scheme requires in fact a non-zero background TKE. Following Masciadri and Jabouille (2001), to produce realistic values of C_N^2 profile in the upper troposphere, this background TKE value has been changed to a preliminary value of $12 \cdot 10^{-4}$ m^2 s^{-2}. The following section describes the calibration of the background value for TKE throughout the atmosphere.

9.2.1 Algorithm Calibration

Masciadri and Jabouille (2001) showed that imposing a constant value for the background value for TKE throughout the atmosphere could result in systematic errors in the optical turbulence prediction. Since it is not physically possible to measure the background value of TKE, the authors proposed various calibration techniques based on the idea to compare the measured and simulated optical turbulence profiles. The following calibration technique (Masciadri and Jabouille, 2001) has been applied to the WRF optical turbulent algorithm.

If M is the total number of nights during which measurements of optical turbulent profiles are available, the calibration technique consists in calculating the profile $a_m(k)$ ($m=1,M$), for each night m, which minimizes the following expression:

$$\sigma_m(k) = \sum_{t=1}^{T} \left[< C_N^2(k)_{obs} > -a_m(k)C_N^2(t,k)_{sim} \right]^2, \qquad (9.5)$$

where k indicates the generic model level. The index t varies from 1 to the number of times T, for each night that the simulated turbulence profiles are available. The $C_N^2(k)_{obs}$ is the observed profile sampled over the model levels. The brackets for the observed profiles indicate the nightly average. The profile $a(k)$ of the correction coefficients averaged over the M nights is then calculated as:

$$a(k) = \frac{\sum_{m=1}^{M} a_m(k)}{M}, \qquad (9.6)$$

Consequently, the profile for the background value of TKE is calculated as follows:

$$E^*_{min}(k) = E_{min}a(k)^{3/2}; \; E_{min} = 12 \cdot 10^{-4} \text{ m}^2 \text{ s}^{-2}$$

This calibration technique only differs from the one proposed by Masciadri and Jabouille (2001) as it calculates the correction coefficients at each model level instead that for a finite number of vertical regions. Naturally, the effectiveness of calibration in making the optical algorithm more robust depends on the observational dataset used. An accurate calibration can be obtained by using a very large sample of nights: the more the sample includes nights/cases representing the range of turbulence behaviors that correspond to the variety of naturally occurring atmospheric conditions, the more effective the calibration will be. This section describes a method for calibrating the optical turbulent algorithm, and the preliminary results indicate that the approach is sound. However, because of the relatively limited size of the observational dataset used in this study, the calibration has not reached its potential, which will require a significantly larger data set.

9.3 Observational Data and Experimental Set-up

The vertical distribution of turbulence over Mauna Kea was measured as a part of a site characterization campaign held during October and December 2002. The Generalized SCIntillation Detection And Ranging (G-SCIDAR) was operated from October 21 to October 24, and from December 12 to December 18, for a total of eleven nights (Tokovinin et al., 2005). The G-SCIDAR is an instrument that remotely measures the vertical distribution of atmospheric turbulence (Vernin and Roddier, 1973; Avila et al., 1997) by analyzing the stellar scintillation images of a binary star target through the turbulent layers present in the atmosphere. All the technical details relative to the G-Scidar are summarized and referenced in Cherubini et al. (2008a; see also Chapter 4).

To validate the optical turbulence algorithm and perform the calibration described in the previous section, the WRF model was rerun for nine nights during the 2002 Mauna Kea monitoring site campaign. The night of the 21st of October and the night of the 18th of December have not been included in the validation/ calibration as the observed data-set collected during those nights is incomplete and therefore the nightly average behavior of turbulence cannot be accurately determined. For each of the nights under investigation, WRF was initialized at 0000 UTC of that same nominal day. The initial and boundary conditions for WRF are the Global Forecasting System (GFS) analyses available every six hours. In an operational framework, a mesoscale model prediction uses a global circulation model analysis as initial conditions and the global model forecasts originating from the analysis as boundary conditions. In an operational framework, the only analysis available at the time the model is run, is the one at the initial time. In the experiments described in this paper the WRF model instead uses the global analyses as boundary conditions. This results in a slightly more accurate model prediction. The use of analyses instead of forecasts for the boundary conditions is chosen to reduce the impact of a potentially inaccurate forecast on the turbulence algorithm validation.

Initially, WRF was run with the default value for $E_{min} = 12 \cdot 10^{-4}$ m^2 s^{-2}. This set of simulations will hereafter be referred to as control runs (CTRL). For each simulation, seeing and C_N^2 are derived from the grid point closest to the summit geographical location. These data are extracted from the WRF domain-3, with horizontal resolution of 1 km, and are available with a 5-minute frequency. Turbulence data from G-SCIDAR were collected between 0600 and 1600 UTC; therefore the model output from the 6th to the 16th hour of WRF forecast is considered in the comparison.

The weather pattern from October 22 to October 24, 2002, was dominated by a zonal upper-level flow and a surface ridge located north of Hawaii. A weak mid-level ridge was strengthening, following passage of a short-wave trough aloft that swept through the central Pacific area a few days earlier (Fig. 9.3). Consequently, the tradewind inversion was gradually strengthening, winds aloft were moderately strong, and substantial shear was present between the summit and the upper troposphere (Fig. 9.4). At the summit, precipitable water ranged between 3 and 5 mm on October 22, and between 2 and 3 mm on the following two nights. Summit wind speeds were in the 2.5-5 m s^{-1} range on October 22 and 5 to 10 m s^{-1} on the following two nights. The atmosphere was gradually stabilizing.

The weather pattern from December 12 to December 17, 2002, was governed by a zonal mid-level ridge ~200 km north of the Big Island that stretched across most of the Eastern Pacific and remained quasi-stationary throughout this period (Fig. 9.5). This pattern helped maintain a strong tradewind inversion near 1500 meters (Fig. 9.6a) and kept the summit-level air mass quite dry with PW values near 1

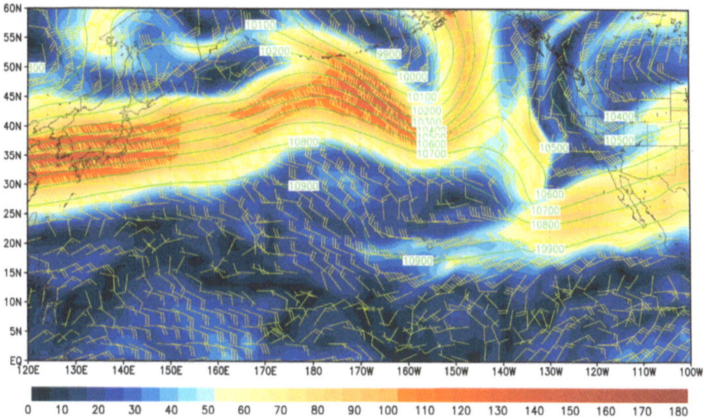

Figure 9.3 Analyses for 0000 UTC 23 Oct 2002 of geopotential height (green lines), wind barbs and wind speed in knots (color shaded) at 250 mb.

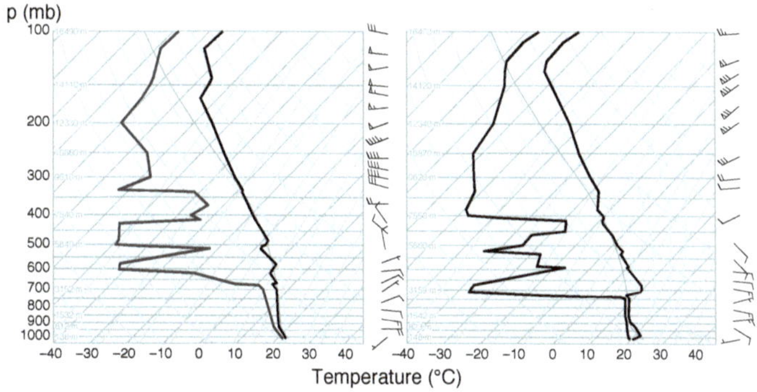

Figure 9.4 Skew-T diagrams for Hilo, HI, at 12 UTC on (a) 22 and (b) 24 Oct 2002.

mm. The position of the ridge also promoted light easterly flow at the summit (5 to 10 m s^{-1}). Flow aloft remained zonal throughout most of this period, and minimal shear existed in the free atmosphere prior to the 18th, as winds aloft remained very weak (Fig. 9.6b). A westerly jet gradually began to drift over the state by 12 UTC on Wednesday Dec. 18, resulting in substantial shear between the summit and the upper-troposphere on Dec. 18 and 19.

9.4 Calibration Results

Figure 9.7a shows the nightly averaged C^2_N observed and predicted profiles (CTRL) for each night. The distribution of the observed profiles shows a larger spread than the distribution of the simulated profiles, suggesting that in the model atmosphere, turbulence production might not be as sensitive to changes in

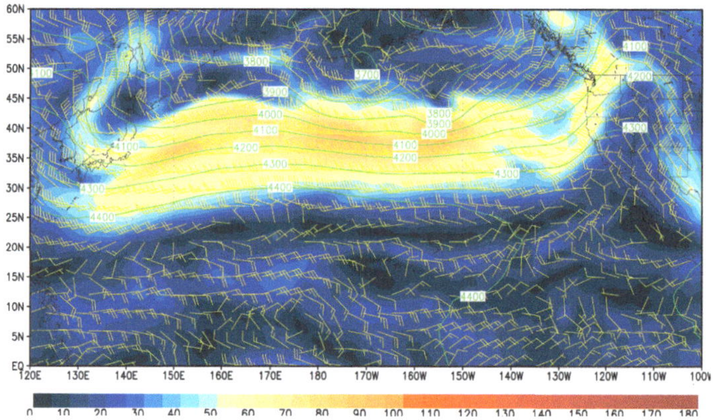

Figure 9.5 Analyses for 0000 UTC 13 Dec 2002 of geopotential height (green lines), wind barbs and wind speed in knots (color shaded) at 600 mb.

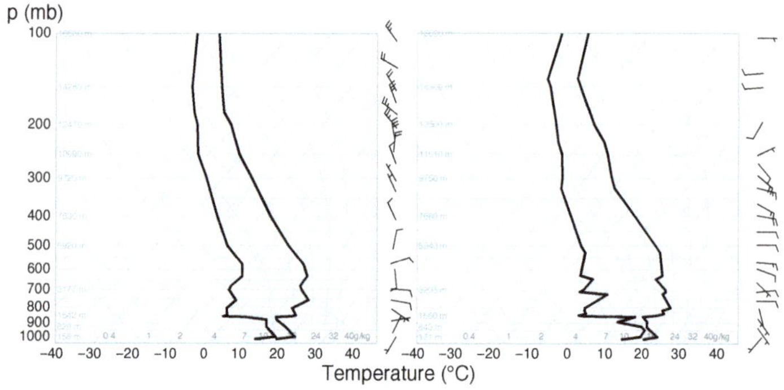

Figure 9.6 Skew-T diagrams for Hilo, HI, at 12 UTC on (a) 13 and (b) 15 Dec 2002.

atmospheric conditions as is in the real atmosphere. Figure 9.7b shows the observed and simulated profiles averaged over the nine nights. The average simulated profile shows a certain degree of underestimation below 13-14 km altitude. This might suggest that the background value for TKE might be smaller than what is needed for this particular site and region of the atmosphere to activate the production of optical turbulence.

The aim of the calibration is to investigate whether values of E_{min} as a function of the altitude might produce better results. For each night the average observed C_N^2 profile has been calculated and sampled over the same model levels. Following the method indicated in Section 9.2.1, the profiles $a_m(k)$ for each night, and the averages over the nine nights $a(k)$, are calculated. The resulting $E_{min}(k)$ is shown in Fig. 9.8 and indicates that higher values for E_{min} should be used in general within

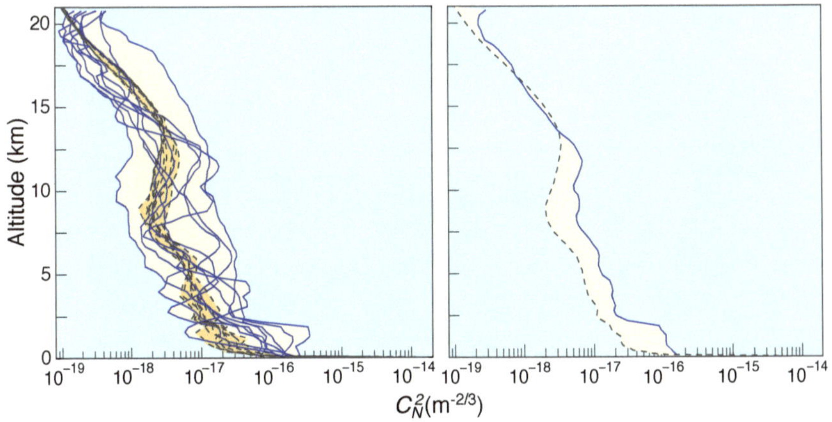

Figure 9.7 a) Average nightly C_N^2 profiles as observed (black solid lines) and predicted (black shaded lines) by the CTRL runs for the nine nights; b) observed (black solid line) and predicted (black shaded lines) CTRL profiles averaged over the nine nights.

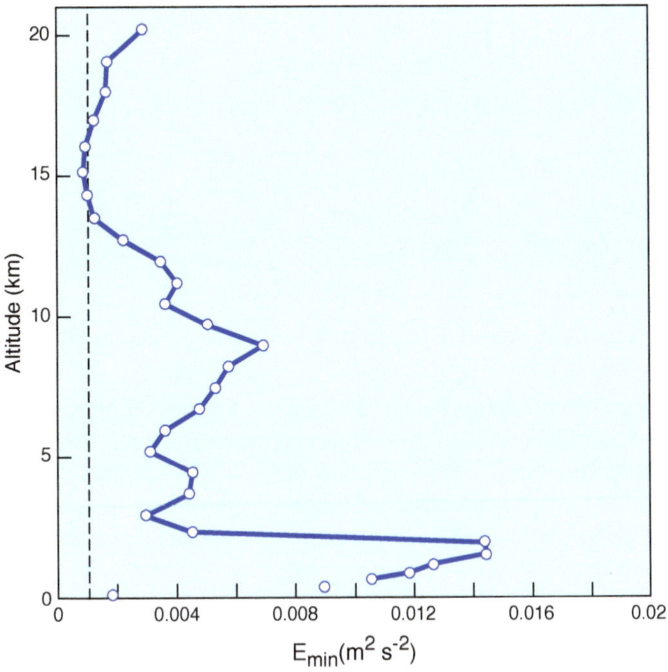

Figure 9.8 Vertical profiles for the background value of TKE, E_{min} after calibration (solid black line). The crosses indicate the model levels height. The dashed line indicates a constant TKE value of $12 \cdot 10^{-4}$ $m^2 s^{-2}$.

the troposphere, and in particular in the atmosphere below 2.5 km and in the upper troposphere, where the average location of the upper part of the jet is expected.

A second set of simulations is then carried out and WRF is run using the variable E_{min} profile within the optical turbulence algorithm. Figures 9.9 and 9.10 show the average simulated profiles for each night for the CTRL runs and after calibration against the average observed profiles. Each plot also shows the wind barbs at each model level at the summit model grid point. The post-calibration simulated profiles reproduce more closely some of the features observed in the measured profiles. For example, the turbulent behavior in the lower atmosphere above the summit (i.e., below 2.5 km) is better captured for most of the nights, particularly for the December portion of the campaign, and for at least one of the nights in October. Also, the turbulent profile in the middle to upper atmosphere (2.5 km $< z <$ 12.5 km) is better represented by the post-calibration profiles than by the CTRLs for all but two nights in December (12/15/02, 12/16/02). Figure 9.10 shows the average over the nine nights for the CTRL, post-calibration, and measured profiles. Not surprisingly the post-calibration profiles more closely reproduce the measured profile, since the calibration method acts to minimize the difference between the single simulated profiles and the nightly averaged one.

Table 9.1 shows that for most nights, there is an improvement in the prediction of seeing by the post-calibration runs. The observed seeing is measured at 85 m above the ground. The values for seeing on the lowest three model levels are listed as the simulated seeing from the two sets of experiments (CTRL and post-calibration). Since the WRF model does not include a model level at 85 m above the ground, the comparison has to be carried out, keeping in mind that the reference for the simulated value falls somewhere in between the second and third model level, which are at heights of 56 m and 112 m, respectively. Consistent with Figs. 9.9 and 9.10, the predicted seeing with post-calibration is in better agreement with the observations, particularly for the nights in December. Less improvement is observed for the first two nights in October.

Overall, an optical turbulence algorithm initialized with a variable profile for E_{min}, produces better prediction in terms of C^2_N profiles and seeing values than what is obtained when using a constant value for E_{min}. Since the role of E_{min} in the turbulence scheme is to numerically allow the turbulence to start developing in very stable conditions, it is reasonable to expect that a variable profile for E_{min} produces better results than a constant profile by allowing the turbulent behavior of the atmosphere to vary.

Although data during the 2002 site-monitoring campaign were collected during periods of stable weather and favorable observing conditions, the weather patterns during the two portions of the campaign resulted in substantial differences in observing conditions, as described earlier. During the October portion of the campaign, strong shear aloft contributed to the optical turbulence and C^2_N profiles show maxima at the entrance and exit of the jet, while the lowest layers of the

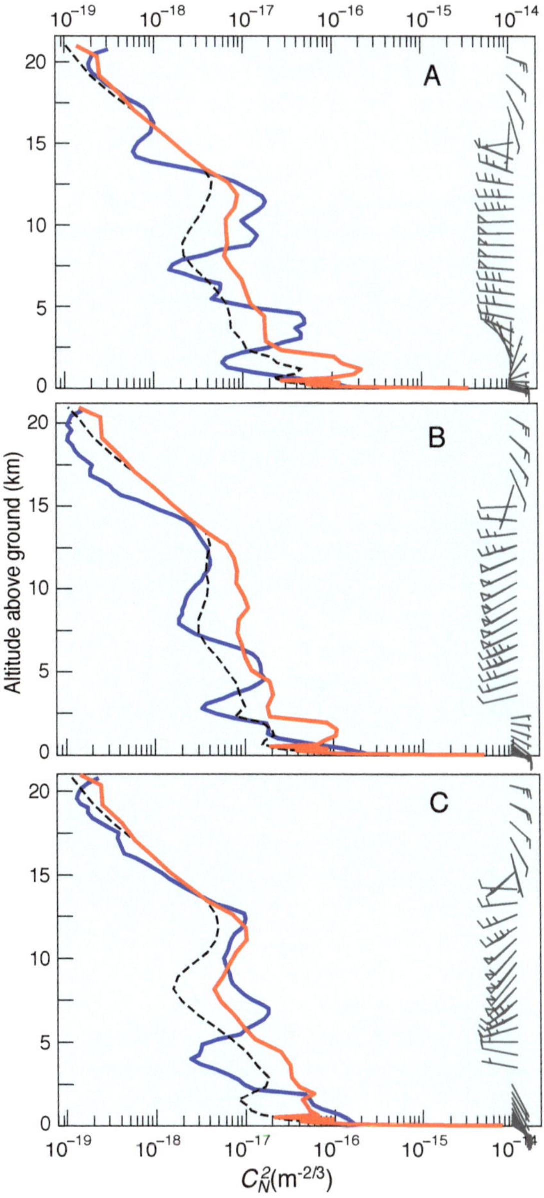

Figure 9.9 Average nightly C_N^2 profiles as observed (black solid line), predicted by the WRF CTRL (black shaded line), and post-calibration (grey solid line) for the (a) 22, (b) 23 and (c) 24 Oct 2002. The average wind vertical profile for each night at the summit grid point is also shown in barbs.

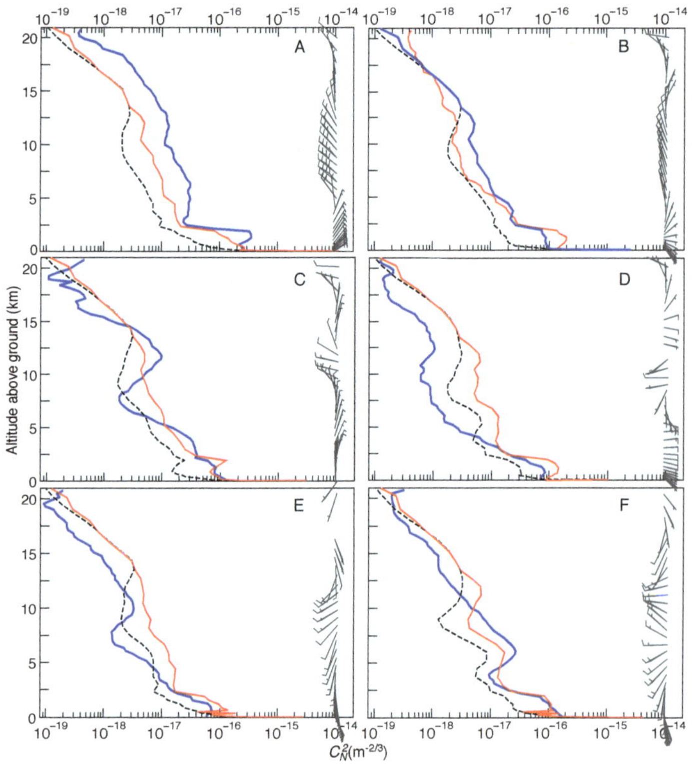

Figure 9.10 Average nightly C_N^2 profiles as observed (black solid line), predicted by the WRF CTRL (black shaded line), and post-calibration (grey solid line) for the (a) 12, (b) 13, (c) 14, (d) 15, (e) 16, and (f) 17 Dec 2002. The average wind vertical profile for each night at the summit grid point is also shown in barbs.

atmosphere do not show large contributions to optical turbulence (Fig. 9.9). In contrast, the C_N^2 profiles for the December portion of the campaign show smaller contributions to the optical turbulence from the upper troposphere, because of the low shear conditions aloft, and a large contribution from the atmosphere below 2.5 km. Because of the different optical turbulence behavior of the atmosphere in these two time frames, it is not surprising that the calibration produces an overall improvement of the WRF optical turbulent algorithm, but discrepancies between the single observed and simulated profiles are still present. It is anticipated that a more accurate calibration can be obtained by using a large sample of nights to better represent the range of turbulence behaviors that correspond to the variety of naturally occurring atmospheric conditions. The goal of this section is to describe in detail a method for calibrating an optical turbulent algorithm. Further refinement of the algorithm calibration is planned at MKWC, using the extensive data sets

Table 9.1 The nightly average values of seeing (arcsec) for each night as measured by the G-Scidar at 85m, and simulated by the CTRL, and the post-calibration experiment for the lowest three WRF model levels.

Observed free seeing (@ 85m)	Simulated (CTRL) $E_{min} = 10 \cdot 12^{-4}$ m^{2}s^{-2}	Simulated Variable E_{min}	Model Level height
	0.54	0.83	16 m
0.54	0.42	0.74	56 m
	0.37	0.71	112 m
	0.59	0.80	16 m
0.46	0.41	0.67	56 m
	0.33	0.61	112 m
	0.73	0.89	16 m
0.57	0.47	0.67	56 m
	0.34	0.57	112 m
	0.84	1.07	16 m
1.15	0.52	0.80	56 m
	0.39	0.71	112 m
	0.46	0.68	16 m
0.76	0.36	0.61	56 m
	0.33	0.59	112 m
	0.46	0.70	16 m
0.63	0.35	0.62	56 m
	0.33	0.61	112 m
	0.39	0.69	16 m
0.44	0.35	0.66	56 m
	0.34	0.65	112 m
	0.47	0.71	16 m
0.43	0.37	0.63	56 m
	0.35	0.62	112 m
	0.63	0.85	16 m
0.68	0.45	0.71	56 m
	0.35	0.64	112 m

collected at Mauna Kea during the Thirty Meter Telescope (TMT) site-monitoring campaign and the Ground Layer Adaptive Optics (GLAO) campaign.

9.5 The Value of Experience: The Human Factor

Synthetic C_N^2 profiles and seeing forecast trends are updated twice daily and posted to the MKWC web pages. Since July 2006 a quantitative seeing forecast is part of the MKWC product stream. The forecaster uses as guidance the output from the implemented algorithm and his/her own experience based on the daily observation of the synoptic and local weather conditions and the reported observing conditions.

Forecaster experience has demonstrated that the two extreme situations of very good and very bad observing conditions can be well anticipated. In fact, excellent seeing is very often correlated to a deep and large-scale ridge parked over the Hawaiian Islands, producing light and uniform winds at and above the summit. In contrast, very bad seeing is well correlated to poor weather conditions at and/or above the summit, usually emanating from a trough to the west of the Island of Hawaii. Both of these large-scale weather patterns are reasonably well handled by numerical weather prediction models, giving the forecaster better guidance from which to construct the final forecast.

The middle part of the seeing spectrum tends to be more difficult to forecast, resulting in greater errors, in part due to the numerous factors that impact the turbulent state of the atmosphere. These factors include, for example, horizontal and vertical wind shear at various levels in the atmosphere induced by upper-level lows, remnant hurricanes/tropical storms, and gravity waves ahead of cold fronts and squall lines, to name a few. Additionally, summit-level winds can stir up boundary layer turbulence, and bring in moisture and/or clouds. Each of these factors, in different ways, has an impact on the observing conditions. None of these factors is easy to quantify just by the human observation. Therefore, guidance from a high-resolution weather prediction model and the embedded optical turbulent scheme represents a very valuable resource in forecast development.

9.6 Conclusions and Future Work

This overview of the implementation and calibration of an optical turbulence algorithm describes part of an operational weather prediction model at the MKWC. The optical turbulence algorithm provides predictions of turbulence profiles and seeing for the summit of Mauna Kea, and therefore guidance to MKWC forecasters.

The model configuration includes a 1-km resolution horizontal grid over the summit of Mauna Kea. In the vertical, the model has forty density-weighted levels, providing finer resolution near the ground (~tenths of meters) and coarser resolution from the tropopause up to the top of the model, which is fixed at 10 mb.

Figure 9.11 C_N^2 profiles averaged over the entire period of the 2002 site-monitoring campaign (nine nights): black line is for the average observed profile, shaded line is the average CTRL profile and grey line is the average calibrated profile.

The algorithm implemented with this model configuration provides an estimate of the turbulence at each model grid point by parameterizing the sub-grid scale optical turbulence processes. Data from the 2002 Mauna Kea site-monitoring campaign were used to validate and calibrate the algorithm.

The results show that the average nightly observed and predicted C^2_N profiles were correlated but the distribution of the measured profiles for the nine nights analyzed shows a larger spread than the distribution of the simulated profiles. This suggests that in the model atmosphere, turbulence production is not as sensitive to changes in atmospheric conditions as it is in the real atmosphere. A comparison of the observed and simulated profiles averaged over the nine nights indicates a certain degree of underestimation of C^2_N below 13-14 km of altitude. Accordingly, the analysis of the average simulated values for the nine nights points to a systematic underestimation when compared to the recorded average seeing values. Following these considerations, a calibration of the background value of TKE, E_{min}, led to the use of a variable profile instead of a constant one. The optical turbulence algorithm initialized with a variable profile for the background value of TKE, E_{min}, produces better predictions in terms of C^2_N profiles and seeing values when compared with the observations. The sample of nine nights used in the calibration is quite small.

A more accurate calibration could be constructed by using a larger sample of nights that more completely represents the range of turbulence behavior associated with the naturally occurring atmospheric variability. The effectiveness of the calibration depends in fact on how close the daily simulated weather pattern is to the patterns that occurred during collection of the observational dataset being modeled. Ideally, a range of calibrations could be constructed for the optical turbulence algorithm that reflects the range of weather patterns that occurred during the observation period. Then different calibrations could, for example, be utilized depending on the present weather pattern. Calibrations could be refined on the basis of seasonal pattern changes (summer ridge vs. winter jet) or synoptic weather types (Kona trough vs. tradewind inversion).

Operationally, each WRF model run takes about 5 h to complete a 60-h forecast. The effect of the implementation of the C_N^2 algorithm within the model on the simulation length is negligible because the algorithm makes use of variables that have already been calculated by the model for other purposes. The seeing and C_N^2 plots available on the MKWC web page are produced in post-processing and they are available at the same time as plots of other meteorological variables.

The great advantage of implementing an optical turbulence algorithm within a weather model is that it allows three-dimensional maps of C_N^2 and seeing to be produced operationally (Fig. 9.12). This output provides graphic guidance for forecasters in daily predictions, and archived data can also be used as a proxy for measurements from an expensive set of instruments in site characterization campaigns.

The MKWC staff learned the importance of the synergy between human experience and model guidance in producing the best products in predicting optical turbulence.

9.6.1 Future Work

At the time of this writing, data are available from the Thirty Meter Telescope (TMT) site-monitoring campaign for the Mauna Kea summit, which extended from July 2005 to May 2008 (TMT Science Advisory Committee, 2007). Further algorithm calibrations will be carried out using these data. Given the large amount of data available it will be possible to calibrate the algorithm and independently validate the algorithm using a subset of the data set that has not been included in the calibration.

A fixed C_N^2 and seeing monitor was installed at the summit of Mauna Kea at the end of the summer of 2009, and data from a Slope Detection and Ranging (SLODAR) instrument able to accurately measure optical turbulence in the lowest atmospheric layers will also be made available to the MKWC. Further algorithm validation and refinement will be performed when data from the new instruments become routinely available.

Figure 9.12 (a) Cross section of C_N^2 at 1200 UTC on 23 October 2002. Cross section latitude is fixed to summit latitude. (b) Aerial map of seeing for the summit of Mauna Kea at 1200 UTC on 23 October 2002. Seeing is plotted on the model level closest to the G-SCIDAR first level (~56 m above ground). (c) as in a) but for 1200 UTC on 13 December 2002. (d) as in b) but for 12 UTC on 13 December 2002.

As computational resources expand in future, the sensitivity of the algorithm performance to the horizontal and vertical resolution of the model will be studied. Numerical instabilities associated with the model atmospheric flow interaction with complex topography will also be investigated.

Authored by
T. Cherubini, S. Businger, and R. Lyman
University of Hawaii, Honolulu, HI

References

References

Abahamid, A., A. Jabiri, J. Vernin, Z. Benkhaldoun, M. Azouit, and A. Agabi, 2004: Optical turbulence modeling in the boundary layer and free atmosphere using instrumented meteorological balloons. *Astron. Astrophys.*, **416**, 1193–1200.

Abahamid, A., J. Vernin, Z. Benkhaldoun, A. Jabiri, M. Azouit, and A. Agabi, 2004: Seeing, outer scale of optical turbulence, and coherence outer scale at different astronomical sites using instruments on meteorological balloons. *Astron. Astrophys.*, **422**(3), 1123–1127.

Agabi, A., J. Borgnino, F. Martin, A. Tokovinin, and A. Ziad, 1995: G.S.M: A grating scale monitor for atmospheric turbulence measurements. II. First measurements of the wavefront outer scale at the O.C.A. *Astron. Astrophys. Suppl.*, **109**, 557–562.

Aussem, A., F. Murtagh, and M. Sarazin, 1994: Dynamical recurrent neural networks and pattern recognition methods for time series prediction: Application to seeing and temperature forecasting in the context of ESO's VLT astronomical weather station. *Vistas Astron.*, **38**, 357–374.

Avila, R., J. Vernin, and E. Masciadri, 1997: Whole atmosphere turbulence profiling with generalized G-SCIDAR. *Appl. Opt.*, **36**, 7898–7905.

Avila, R., A. Ziad, J. Borgnino, F. Martin, A. Agabi, and A. Tokovinin, 1997: Theoretical spatiotemporal analysis of angle of arrival induced by atmospheric turbulence as observed with the grating scale monitor experiment. *J. Optic. Soc. Am. A*, **14**(11), 3070–3082.

Avila, R., and J. Vernin, 1999: Mechanism of formation of atmospheric turbulence relevant for optical astronomy. *Interstellar Turbulence*, J. Franco and A. Carraminana, Eds., Cambridge University Press, UK, 5–11.

Avila, R., E. Masciadri, J. Vernin, and L. Sánchez, 2004: Generalized SCIDAR measurements at San Pedro Mártir. I. Turbulence profile statistics. *Publ. Astron. Soc. Pac.*, **116**(821), 682–692.

Avila, R., E. Carrasco, F. Ibanez, J. Vernin, J.-L. Prieur, and D. X. Cruz, 2006: Generalized SCIDAR measurements at San Pedro Mártir. II. Wind profile statistics. *Publ. Astron. Soc. Pac.*, **118**(841), 503–515.

Avila, R., 2007: Low Layer SCIDAR: a turbulence profiler for the first kilometer with very high altitude-resolution. *Proc. Symp. on Seeing*, T. Cherubini and S. Businger, Eds., Kona, Hawaii.

Avila, R., J. L. Aviles, R. Wilson, M. Chun, T. Butterley, and E. Carrasco, 2008: LOLAS: an optical turbulence profiler in the atmospheric boundary layer with extreme altitude resolution. *Mon. Not. R. Astron. Soc.*, **387**(4), 1511–1516.

Azouit, M., and J. Vernin, 2005: Optical turbulence profiling with balloons relevant to astronomy and atmospheric physics. *Publ. Astron. Soc. Pac.*, **117**(831), 536–543.

Babcock, H. W., 1963: The possibility of compensating astronomical seeing. *Pub. Astron. Soc. Pac.*, **65**, 229–236.

Barletti, R., G. Ceppatelli, E. Moroder, L. Paterno, and A. Righini, 1974: High resolution optical astronomy and atmospheric microturbulence. *Rivista Italiana di Geofisica*, **23**, 264–266.

Barletti, R., G. Ceppattelli, L. Paternò, A. Righini, and N. Speroni, 1976: Mean vertical profile of atmospheric turbulence relevant for astronomical seeing. *J. Opt. Soc. Amer.*, **66**(12), 1380–1383.

Barletti, R., G. Ceppatelli, L. Paternò, A. Righini, and N. Speroni, 1977: Astronomical site testing with balloon borne radiosondes—Results about atmospheric turbulence, solar seeing and stellar scintillation. *Astron. Astrophys.*, **54**(3), 649–659.

Beckers, J., 1988: Very large telescopes and their instrumentation, *Proc. ESO Conf. on Very Large Telescopes and their Instrumentation*, Garching, European Southern Observatory, 693.

Beckers, J., 2001: A seeing monitor for solar and other extended object observations. *Exp. Astron.*, **12**, 1–20.

Beckers, J. M., 1993: Adaptive optics for astronomy: principles, performance, and applications. *Ann. Rev. Astron. Astrophys.*, **31**, 13–62.

Bedulin, A. N., et al., 1991: Intercomparisons of fast response micrometers and acoustic sounders (of Institute of Atmospheric Physics—IAP, USSR Academy of Sciences) at Mt. Maydanak. Internal report, Sternberg Astronomical Institute of Moscow University, 50 pp.

Beland, R., 1993: Propagation through atmospheric optical turbulence. *Atmospheric Propagation of Radiation,* F. G. Smith, Ed., Vol. 2, *The Infrared and Electro-Optical Systems Handbook*, J. Accetta and D. Shumaker, Ex. Eds., Infrared Information Analysis Center, Ann Arbor, MI and SPIE Optical Engineering Press, Bellingham WA, 217–224.

Bougeault, P., and P. Lacarrere, 1989: Parameterization of orography-induced turbulence in a meso-beta-scale model. *Mon. Wea. Rev.*, **117**, 1872–1890.

Bougeault, P., C. De Hui, B. Fleury, and J. Laurent, 1995: Investigation of seeing by means of an atmospheric mesoscale numerical simulation. *Appl. Optics*, **34**, 3481–3488.

Britton, M. C., 2006: The anisoplanatic point-spread function in adaptive optics. *Publ. Astron. Soc. Pac.*, **118**, 885–900.

Brown, E. H., and F. F. Hall, 1978: Advances in atmospheric acoustics. *Rev. Geophys.*, **16**(1), 47–110.

Brown, J. H., R. E. Good, P. M. Bench, and G. Faucher, 1982: Sonde measurement for comparative measurements of optical turbulence. Air Force Geophys. Lab., AFGL-TR-82-0079, ADA118740, 46 pp.

Brown, J. H., and R. R. Beland, 1988: A site comparison of optical turbulence in the lower stratosphere at night using thermosonde data. *Physica Scripta.*, **37**(3), 424–426.

Buffa, F., and I. Porceddu, 1997:Temperature forecast and dome seeing minimization—I. A case study using a neural network model. *Astron. Astrophys. Suppl. Ser.*, **126**(3), 547–553.

Bufton, J. L., 1973a: Correlation of microthermal turbulence data with meteorological soundings in the troposphere. *J. Atmos. Sci.*, **30**, 83–87.

Bufton, J. L., 1973b: Comparison of vertical profile turbulence structure with stellar observations. *Appl. Optics*, **12**, 1785–1793.

Bufton, J. L., 1975: A radiosonde thermal sensor technique for measurement of atmospheric turbulence. NASA Technical Report, Goddard Space Flight Ctr., NASA TN D-7867.

Businger, S., R. McLaren, R. Ogasawara, D. Simons, and R. J. Wainscoat, 2002: Starcasting. *Bull. Amer. Meteor. Soc.*, **83**, 858–871.

Butterley T., R. Wilson, and M. Sarazin, 2006: Determination of the profile of atmospheric optical turbulence strength from SLODAR data. *Mon. Not. R. Astron. Soc.*, **369**, 835–845.

Caccia, J. L., M. Azouit, and J. Vernin, 1987: Wind and C2N profiling by single-star scintillation analysis. *Appl. Opt.*, **26**, 1288–1294.

Chernov, L. A., 1960: *Wave Propagation in Random Media*. Dover, New York.

Cherubini, T., S. Businger, C. Velden, and R. Ogasawara, 2006: The impact of satellite-derived atmospheric motion vectors on mesoscale forecasts over Hawaii. *Mon. Wea. Rev.*, **134**, 2009–2020.

Cherubini, T., S. Businger, R. Lyman, and M. Chun, 2008: Modeling optical turbulence and seeing over Mauna Kea. *J. Appl. Meteorol.*, **47**, 1140–1155.

Chun, M., R. Wilson, R. Avila, and D. Weir, 2007: The Mauna Kea ground-layer characterization campaign. *Proc. Symp. on Seeing*, T. Cherubini and S. Businger, Eds., Kona, Hawaii.

Conan, R., 2000: Modélisation des effets de l'échelle externe de cohérence spatiale du front d'onde pour l'observation à haute résolution angulaire en astronomie. Ph.D. dissertation, Nice University, Nice, France.

Coulman, C. E., 1985: Fundamental and applied aspects of astronomical "seeing". *Ann. Rev. Astr. Astroph.*, **23**, 19–57.

Coulman, C. E., J. Vernin, Y. Coqueugniot, and J. L. Caccia, 1988: Outer scale of turbulence appropriate to modeling refractive-index structure profiles. *Appl. Opt.*, **27**(1), 155–160.

Coulman, C. E., J. Vernin, and A. Fuchs, 1995: Optical seeing-mechanism of formation of thin turbulent laminae in the atmosphere. *Appl. Opt.*, **34**(24), 5461–5474.

Cuxart, J., P. Bougeault, and J.-L. Redelsperger, 1995: Turbulence closure for a non-mydrostatic model. *Proc. 11th Symp. Boundary Layer Turb.*, Charlotte, NC, Amer. Meteor. Soc., 409.

Dainty, J. C., and R. J. Scaddan, 1975: Measurements of the atmospheric transfer function at Mauna Kea, Hawaii. *Mon. Not. R. Astron. Soc.*, **170**, 519–532.

Dalaudier, F., C. Sidi, M. Crochet, and J. Vernin, 1994: Direct evidence of 'sheets' in the atmospheric temperature field. *J. Atmos. Sci.*, **51**, 237–248.

Danilov, S. D., A. E. Guryanov, M. A. Kallistratova, I. V. Petenko, S. P. Singal, D. R. Pahwa, and B. S. Gera, 1992: Acoustic calibration of sodars. *Meas. Sci. Technol.*, **3**(10), 1001.

De Young, D., and R. D. Charles, 1995: Numerical simulation of airflow over potential telescope sites. *Astron. J.*, **110**(6), 3107.

Eaton, F. D., G. D. Nastrom, D. T. Kyrazis, D. G. Black, W. T. Black, and R. A. Black, 2009: Preliminary VHF radar and high-data-rate optical turbulence profile observations using a balloon-ring platform. *Proc. SPIE*, 7463, 746309.

Echevarria, J., and Coauthors, 1998: Site testing at Observatorio Astronómico Nacional in San Pedro Mártir. *Rev. Mex. Astron. Astr.*, **34**, 47-60.

Egner, S., and E. Masciadri, 2007: A G-SCIDAR for ground-layer turbulence measurements at high vertical resolution. *Publ. Astron. Soc. Pac.*, **119**(862), 1441–1448.

Egner, S., E. Masciadri, and D. McKenna, 2007: Generalized SCIDAR measurements at Mount Graham. *Publ. Astron. Soc. Pac.*, **119**(856), 669–686.

Els, S. G., and Coauthors, 2008: Study on the precision of the multiaperture scintillation sensor turbulence profiler (MASS) employed in the site testing campaign for the thirty meter telescope. *Appl. Opt.*, **47**(14), 2610–2618.

Erasmus, D. A., and B. C. Barnes, 1989: Atmospheric boundary layer conditions at three Mauna Kea sites. Institute for Astronomy, University of Hawaii, Technical Report, 53pp.

Foy, R., and A. Laberie, 1985: Feasibility of adaptive telescope with laser probe. *Astron. Astrophys.*, **152**(2), L29–L31.

Fried, D. L., 1965: Statistics of a geometric representation of wavefront distortion. *J. Opt. Soc. Amer.*, **55**(11), 1427–1431.

Fried, D. L., 1966: Optical resolution through a randomly inhomogeneous medium for very long and very short exposures. *J. Opt. Soc. Amer.*, **56**(10), 1372–1379.

Fried, D. L., 1979: Angular dependence of the atmospheric turbulence effect in speckle interferometry. *Opt. Acta*, **26**(5), 597–613.

Fuchs, A., M. Tallon, and J. Vernin, 1998: Focusing on a turbulent layer: Principle of the "generalized SCIDAR". *Publ. Astron. Soc. Pac.*, **110**(743), 86–91.

Fusco, T., and Coauthors, 2004: NAOS on-line characterization of turbulence parameters and adaptive optics performance. *J. Opt. A*, **6**(6), 585.

Gallée, H., and G. Schayes, 1994: Development of a three-dimensional meso-gamma; primitive equations model. *Mon. Wea. Rev.*, **122**, 671–685.

Gavrilov, N. M., H. Luce, M. Crochet, F. Dalaudier, and S. Fukao, 2005: Turbulence parameter estimations from high-resolution balloon temperature measurements of the MUTSI-2000 campaign. *Ann. Geophys.*, **23**, 2401–2413.

Geissler, K., and E. Masciadri, 2006: Meteorological parameter analysis above Dome C using data from the European Centre for Medium-Range Weather Forecasts. *Publ. Astron. Soc. Pac.*, **118**, 1048–1065.

Gill, A., 1982: *Atmosphere Ocean Dynamics*. Academic Press, London, 662 pp.

Gillingham, P. R., 1978: AAT dome seeing. Progress Report, Anglo-Australian Observatory, Epping.

Goodwin, M., C. Jenkins, and A. Lambert, 2007: Improved detection of atmospheric turbulence with SLODAR. *Opt. Exp.*, **15**, 14844–14860.

Grell, G. A., J. Dudhia, and D. R. Stauffer, 1985: A description of the fifth-generation PENN State/NCAR mesoscale model (MM5). NCAR/TN-398+STR. Technical report, National Center for Atmospheric Research, Boulder, CO.

Habib, A., J. Vernin, Z. Benkhaldoun, and H. Lanteri, 2006: Single star SCIDAR: atmospheric parameters profiling using the simulated annealing alogorithm. *Mon. Not. R. Astron. Soc.*, **368**, 1456–1462.

Hardy, J. W., 1998: *Adaptive Optics for Astronomical Telescopes*. Oxford University Press, New York, NY, 448 pp.

Harlan, E. A., and M. F. Walker, 1965: A star-trail telescope for astronomical site-testing. *Pub. Astron. Soc. Pac.*, **77**, 246–252.

Harris, C. M., 1966: Absorption of sound in air versus humidity and temperature. *J. Acoust. Soc. Am.*, **40**(1),148–159.

Haugen, D. A., and J. C. Kaimal, 1978: Measuring temperature structure parameter profiles with an acoustic sounder. *J. Appl. Meteorol.*, **17**, 895–899.

Hickson, P., and K. Lanzetta, 2004: Measuring atmospheric turbulence with a lunar scintillometer array. *Publ. Astron. Soc. Pac.*, **116**, 1143–1152.

Hill, R. J., 1978: Models of the scalar spectrum for turbulent advection. *J. Fluid Mech.*, **88**(3), 541–562.

Hong, S.-Y., J. Dudhia, and S.-H. Chen, 2004: A revised approach to ice microphysical processes for the bulk parameterization of clouds and precipitation. *Mon. Wea. Rev.*, **132**, 103–120.

Hosfeld, R., 1954: Comparisons of stellar scintillation with image motion. *J. Opt. Soc. Am.*, **44**(4), 284–287, doi:10.1364/JOSA.44.000284.

Hufnagel, R. E., 1974: Optical propagation through turbulence. OSA Technical Digest Series Paper, WA1, Opt. Soc. Am., Washington, DC.

Janjic, Z. I., 2002: Nonsingular implementation of the Mellor–Yamada level 2.5 scheme in the NCEP meso model. NCEP Office Note, No. 437, 61 pp.

Jumper, G. Y., H. M. Polchlopek, R. R. Beland, E. A. Murphy, P. Tracy, and K. Robinson, 1997: Balloon-borne measurements of atmospheric temperature fluctuations. *Proc. 28th Plasmadynamics and Lasers Conf.*, Atlanta, GA, AIAA-97-2353.

Jumper, G. Y., R. R. Beland, and P. Tracy, 1999: Investigating sources of error in balloon-borne optical turbulence measurements. *Proc. 30th Plasmadynamics and Lasers Conf.*, Norfolk, VA, AIAA 99-03618.

Jumper, G. Y., E. A. Murphy, D. J. Strauss, and R. R. Beland, 2002: Using temperature to detect wake contamination of thermosonde data. *Proc. 33rd Plasmadynamics and Lasers Conference*, Maui, HI, AIAA-2002-2275.

Jumper, G. Y., J. Vernin, M. Azouit, and H. Trinquet, 2005: Comparison of recent measurements of atmospheric optical turbulence. *Proc. 36th Plasmadynamics and Lasers Conf.*, Toronto, Canada, AIAA-2005-4778.

Jumper, G. Y., E. A. Murphy, F. H. Ruggiero, J. R. Roadcap, A. J. Ratkowski, J. Vernin, and H. Trinquet, 2007: OHP-APT 2002 gravity wave campaign: waves, turbulence and forecasts. *Environ. Fluid Mech.*, 7(5), 351–370.

Kain, J. S., and J. M. Fritsch, 1990: A one-dimensional entraining/detraining plume model and its application in convective parameterization. *J. Atmos. Sci.*, **47**, 2784–2802.

Kain, J. S., and J. M. Fritsch, 1993: Convective parameterization for mesoscale models: The Kain-Fritcsh scheme. In *The Representation of Cumulus Convection in Numerical Models*, K. A. Emanuel and D. J. Raymond, Eds., Amer. Meteor. Soc., 246 pp.

Kellerer, A., and Tokovinin, A., 2007: Atmospheric coherence time in interferometry: definition and measurement. *Astron. Astrophys.*, **461**, 775–781.

Klemp, J. B., W. C. Skamarock, and J. Dudhia, 2007: Conservative split-explicit time integration methods for the compressible nonhydrostatic equations. *Mon. Wea. Rev.*, **135**, 2897–2913.

Klueckers, V. A., N. J. Wooder, T. W. Nicholls, M. J. Adcock, I. Munro, and J. C. Dainty, 1998: Profiling of atmospheric turbulence strength and velocity using a generalized SCIDAR technique. *Astron. Astrophys. Suppl.*, **130**, 141–155.

Kolmogorov, A., 1941: The local structure of turbulence in incompressible viscous fluid for very large Reynolds' numbers (in Russian). *Akademiia Nauk SSSR Doklady*, **30**(4), 301–305.

Kolmogorov, A. N., 1991: The local structure of turbulence in incompressible viscous fluid for very large Reynolds Numbers. *Proc. Roy. Soc. London A*, **434**, 1890, 9–13. First published by Dokl. Akad. Nauk SSSR, 30(4) (1941).

Kornilov, V., A. Tokovinin, O. Voziakova, A. Zaitsev, N. Shatsky, S. Potanin, and M. Sarazin, 2003: MASS: a monitor of the vertical turbulence distribution. *Proc. SPIE*, 4839, 837–845.

Kornilov, V., A. Tokovinin, N. Shatsky, O. Voziakova, S. Potatin, and B. Safonov, 2007: Combined MASS-DIMM instrument for atmospheric turbulence studies. *Mon. Not. R. Astron. Soc.*, **383**, 1268–1278.

Lafore, J-P., and Coauthors, 1998: The meso-Nh atmospheric simulation system. Part I: adiabatic formulation and control simulations. *Ann. Geophys.*, **16**, 90–109.

Lascaux, F., E. Masciadri, S. Hagelin, and J. Stoesz, 2008: Meso-Nh simulations of the atmospheric flow above the Internal Antarctic Plateau. *Proc. SPIE*, 7012, 70124D, doi:10.1117/12.787571.

Lascaux, F., E. Masciadri, S. Hagelin, and J. Stoesz, 2010: Optical turbulence vertical distribution with standard and high resolution at Mt. Graham. *Mon. Not. Roy. Astron. Soc.*, to be submitted.

Lawrence, J. S., M. C. B. Ashley, A. Tokovinin, and T. Travouillon, 2004: Exceptional astronomical seeing conditions above Dome C in Antarctica. *Nature*, **431**(7006), 278–281.

LeLouarn, M., and N. Hubin, 2006: Improving the seeing with wide-field adaptive optics in the near-infrared. *Mon. Not. R. Astron. Soc.*, **365**(4), 1324–1332.

Lombardi, G., J. Navarrete, and M. Sarazin, 2008: Combining turbulence profiles from MASS and SLODAR: A Study of the evolution of the seeing at Parnal. E-ELT Programme, E-ELT-TRE-222-0215, issue 1, European Organization for Astronomical Research in the Southern Hemisphere, 15 pp.

Maire, J., A. Ziad, J. Borgnino, and F. Martin, 2007: Measurements of profiles of the wavefront outer scale using observations of the limb of the Moon. *Mon. Not. R. Astron. Soc.*, **377**(3), 1236–1244.

Mariotti, J. M., and G. P. Di Benedetto, 1984: Path length stability of synthetic aperture telescopes—The case of the 25 CM CERGA interferometer. *IAU Colloq. 79: Very Large Telescopes, Their Instrumentation and Programs*, M.-H. Ulrich and K. Kjaer, Eds., 257–265.

Mariotti, J. M., 1994: Adaptive optics for long baseline optical interferometry. *Adaptive Optics for Astronomy*, D. Alloin and J. M. Mariotti, Eds., Vol. 1993, Kluwer Academic Publishers, 309–320.

Martin, F., A. Tokovinin, A. Ziad, R. Conan, J. Borgnino, R. Avila, A. Agabi, and M. Sarazin, 1998: First statistical data on wavefront outer scale at La Silla observatory from the GSM instrument. *Astron. Astrophys.*, **336**, L49–L52.

Martin, F., R. Conan, A. Tokovinin, A. Ziad, H. Trinquet, J. Borgnino, A. Agabi, and M. Sarazin, 2000: Optical parameters relevant for High Angular Resolution at Paranal from GSM instrument and surface layer contribution. *Astron. Astrophys. Suppl.*, **144**(1), 39–44.

Martin, H. M., 1987: Image motion as a measure of seeing quality. *Publ. Astron. Soc. Pac.*, **99**, 1360–1370.

Masciadri, E., and J. Vernin, 1997: Optical technique for inner-scale measurement: possible astronomical applications. *Appl. Opt.*, **36**(6), 1320.

Masciadri, E., 1998. Astronomie et haute resolutione angulaire. Ph.D. dissertation, Dept. d'Astrophysique, Universite de Nice, Nice, France.

Masciadri, E., J. Vernin, and P. Bougeault, 1999a: 3D mapping of optical turbulence using an atmospheric numerical model. I: A useful tool for the ground-based astronomy. *Astron. Astrophys. Suppl. Ser.*, **137**, 185–202.

Masciadri, E., J. Vernin, and P. Bougeault, 1999b: 3D mapping of optical turbulence using an atmospheric numerical model. II: First results at Cerro Paranal. *Astron. Astrophys. Suppl. Ser.*, **137**, 203–216.

Masciadri, E., 2000: Astronomical sites evaluations in the visible and radio range. Invited talk, Marrakech, November 13–17, 2000.

Masciadri, E., and T. Garfias, 2001: Wavefront coherence time seasonal variability and forecasting at the San Pedro Mártir site. *Astron. Astrophys.*, **366**(2), 708–716.

Masciadri, E., and P. Jabouille, 2001: Improvements in the optical turbulence parameterization for 3D simulations in a region around a telescope. *Astron. Astrophys.*, **376**(2), 727–734.

Masciadri, E., J. Vernin, and P. Bougeault, 2001: 3D numerical simulations of optical turbulence at the Roque de Los Muchachos Observatory using the atmospherical model Meso-Nh. *Astron. Astrophys.*, **365**(3), 699–708.

Masciadri, E., R. Avila, and L. J. Sánchez, 2002: First evidence of the finite horizontal extent of the optical turbulence layers. Implications for new adaptive optics techniques. *Astron. Astrophys.*, **382**(1), 378–388.

Masciadri, E., 2003: Near ground wind simulations by a mesoscale atmospherical model for the ELTs site selection. *Rev. Mex. Astron. Astr.*, **39**, 249–259.

Masciadri, E., R. Avila, and L. J. Sánchez, 2004: Statistic reliability of the Meso-Nh atmospherical model for 3D CN 2 simulations. *Rev. Mex. Astron. Astr.*, **40**, 3–14.

Masciadri, E., and S. Egner, 2006: First seasonal study of the optical turbulence with an atmospheric model. *Publ. Astron. Soc. Pac.*, **118**(849), 1604–1619.

McGinley, J. A., 1989: The local analysis and prediction system. *Proc. 12th Conf. on Analysis and Forecasting*, Monterey, CA, Amer. Meteor. Soc., 15–20.

McHugh, J. P., G. Y. Jumper, and M. Chun, 2008: Balloon thermosonde measurements over Mauna Kea, and comparison with seeing measurements. *Publ. Astron. Soc. Pac.*, **120**, 1318–1324.

Michel, R., J. Echevarria, R. Costero, O. Harris, J. Magallón, and K. Escalante, 2003: Seeing measurements at San Pedro Martir Observatory using the DIMM method. *Rev. Mex. Astron. Astr.*, **39**, 291.

Mlawer, E. J., S. J. Taubman, P. D. Brown, M. J. Iacono, and S. A. Clough, 1997: Radiative transfer for inhomogeneous atmosphere: RRTM, a validated correlated-k model for the longwave. *J. Geophys. Res.*, **102** (D14), 16663–16682.

Murphy, E. A., P. Tracy, R. R. Beland, G. Y. Jumper, K. Robinson, and G. Clement, 2007: Thermosonde 2007. AFRL-RV-AJ-TR-2007-1129, AFRL/VSBYA, Hanscom AFB, MA.

Murtagh, F., and M. Sarazin, 1993: Nowcasting astronomical seeing: a study of ESO La Silla and Paranal. *Publ. Astron. Soc. Pac.*, **105**, 932–939.

Neff, W. D, 1975: Quantitative evaluation of acoustic echoes from the planetary boundary layer. NOAA Technical Report to the U.S. Dept. of Commerce, NOAA TR ERL 322-WPL 38, 34 pp.

Obukhov, A. M., 1949: Structure of the temperature field in a turbulent flow (in Russian). *Izv. Akad. Nauk SSSR Ser. Geogr. i Geopfiz*, **13**, 58–69.

Ochs, G. R., T. Wang, R. S. Lawrence, and S. F. Clifford, 1976: Refractive-turbulence profiles measured by one-dimensional spatial filtering of scintillations. *Appl. Opt.*, **15**(10), 2504–2510.

Otten, L. J., A. Jones, D. Black, J. Lane, R. Hugo, J. Beyer, and M. C. Roggeman, 2000: Precision tropopause turbulence measurements. *Proc. SPIE*, 4125, 33–40.

Ottersten, H., 1969: Atmospheric structure and radar backscattering in clean air. *Radio Sci.*, **4**, 1179–1193.

Persson, S. E., D. M. Carr, and J. H. Jacobs, 1990: Las Campanas observatory seeing measurements. *Exp. Astr.*, **1**(3), 195–212.

Peskoff, A., 1968: Theory of remote sensing of clear-air turbulence profiles. *J. Opt. Soc. Am.*, **58**(8), 1032–1037.

Prieur, J. L., G. Daigne, and R. Avila, 2001: SCIDAR measurements at Pic du Midi. *Astron. Astrophys.*, **371**, 366–377.

Prieur, J. L., R. Avila, G. Daigne, and J. Vernin, 2004: Automatic determination of wind profiles with generalized SCIDAR. *Publ. Astron. Soc. Pac.*, **116**(822), 778–789.

Racine, R., D. Salmon, D. Cowley, and J. Sovka, 1991: Mirror, dome, and natural seeing at CFHT. *Publ. Astron. Soc. Pac.*, **103**(1020), 26.

Racine, R., 2005: Altitude, elevation, and seeing. *Publ. Astron. Soc. Pac.*, **117**(830), 401–410.

Ragazzoni, R., E. Marchetti, and P. Valente, 2000: Adaptive-optics corrections available for the whole sky. *Nature*, **403**, 54–56.

Raman, S., 1977: The observed generation and breaking of atmospheric internal gravity waves over ocean. *Bound. Layer Meteorol.*, **12**, 331–349.

Raman, S., 1982: Dynamics of the atmospheric boundary layer during the 1980 total solar eclipse. *Proc. Indian National Science Academy*, 48, 187–195.

Raman, S., C. Nagle, and G. S. Raynor, 1982: Seasonal variations in the formation of internal gravity waves at a coastal site. *J. Appl. Meteorol.*, **21**, 237–242.

Raman, S., P. Boone, and K. S. Rao, 1990: Observations and numerical simulation of the evolution of the tropical planetary boundary layer during total solar eclipses. *Atmos. Environ.*, **24A**, 789–799.

194

Redelsperger, J., and G. Sommeria, 1981: Méthode de représentation de la turbulence d'échelle inférieure à la maille pour un modèle tri-dimensionnel de convection nuageuse. *Bound. Layer Meteor.*, **21**, 509–530.

Reynolds, O., 1883: An experimental investigation of the circumstances which determine whether the motion of water shall be direct or sinuous, and of the law of resistance in parallel channels. *Phil. Trans. Roy. Soc.*, **174**, 935–982.

Rigaut, F., B. Ellerbroek, and R. Flicker, 2000: Principles, limitations, and performance of multiconjugate adaptive optics. *Proc. SPIE*, 4007, 1022–1031.

Rigaut F., 2002: Ground conjugate wide field adaptive optics for the ELTs. Beyond Conventional *Adaptive Optics, Proc. 3D Optical Turbulence Characterization for the New Class of Adaptive Optics Techniques, ESO Conf. and Workshop*, E. Vernet, R. Ragazzoni, S. Esposito, and N. Hubin, Eds., Vol. 58, Garching, Germany, 11–16.

Roadcap, J. R., and E. A. Murphy, 1999: Comparison of isoplanatic angles derived from thermosonde and optical measurements. *Pure Appl. Geophys.*, **156**(3), 503–524.

Rocca, A., F. Roddier, and J. Vernin, 1974: Detection of atmospheric turbulent layers by spatiotemporal and spatioangular correlation measurements of stellar-light scintillation. *J. Opt. Soc. Am.*, **64**(7), 1000–1004.

Roddier, C., and F. Roddier, 1973: Correlation measurements on the complex amplitude of stellar plane waves perturbed by atmospheric turbulence. *J. Opt. Soc. Am.*, **63**(6), 661–663.

Roddier, F., 1981: The effects of atmospheric turbulence in optical astronomy. *Progress in Optics*, E. Wolf, Ed., North-Holland Publishing, Vol. 19, 281–376.

Roddier, F., J.-M. Gilli, and G. Lund, 1982: On the origin of speckle boiling and its effects in stellar speckle interferometry, *J. Opt. (Paris)*, **13**, 263–271.

Roddier, F., and J. P. Lena, 1984: Long baseline Michelson Interferometry with large ground-based telescopes operating at optical wavelength. *J. Optics Paris*, **15**, 171–182.

Roddier, F., Ed., 1999: *Adaptive Optics in Astronomy*. Cambridge Univ. Press, UK, 420 pp.

Sarazin, M., and F. Roddier, 1990: The ESO differential image motion monitor. *Astron. Astrophys.*, **277**, 294–300.

Sarazin, M., T. Butterley, A. Tokovinin, T. Travouillon, and R. Wilson, 2005: The Tololo SLODAR campaign. [Available online http://www.eso.org/gen-fac/pubs/astclim/paranal/asm/slodar/The_Tololo_SLODAR_Campaign.htm.]

Sarazin, M., T. Butterley, J. Navarrete, and R. Wilson, 2007: Assembling composite vertical atmospheric turbulence profiles from DIMM, SLODAR and MASS contemporaneous records at Paranal. *Proc. Symp. on Seeing*, T. Cherubini and S. Businger, Eds., Kona, Hawaii.

Sasiela, R. J., 1994: *Electromagnetic Wave Propagation in Turbulence*. Springer-Verlag, Berlin.

Scheglov, P. V., 1980: *Astron. Tsirk.*, **1124**, 3 (in Russian).Schöck, M., D. Le Mignant, G. A. Chanan, and P. L. Wizinowich, 2003: Atmospheric turbulence characterization with the Keck adaptive optics systems. *Proc. SPIE*, 4839, 813.

Schoeck, M., and Coauthors, 2008: TMT site testing survey: Calibration and results. *Optical Turbulence—Astronomy Meets Meteorology*, E. Masciadri and M. Sarazin, Eds., Nymphes Bay, Porto Conte, Sardinia, Imperial College Press, 98-107.

Simpson, M. D., and S. Raman, 2004: Role of the land plume in the transport of ozone over the ocean during INDOEX (1999). *Bound. Layer Meteorol.*, **111**, 133–152.

Socas-Navarro, H., J. Beckers, P. Brandt, et al., 2005: Solar site survey for the advanced technology solar telescope. I. Analysis of the seeing data. *Pub. Astron. Soc. Pac.*, **117**(837), 1296–1305.

Stock, J., and G. Keller, 1961: Astronomical seeing. *Stars and Stellar Systems*, G. P. Kuiper and B. M. Middlehurst, Eds., Chicago Univ. Press, Chicago, Vol. 1., 138.

Swain, M. R., and H. Gallée, 2006: Antarctic boundary layer seeing. *Publ. Astron. Soc. Pac.*, **118**(846), 1190–1197.

Tallon, M., and R. Foy, 1990: Adaptive telescope with laser probe: isoplanatism and cone effect. *Astron. Astrophys.*, **235**, 549–557.

Tatarski, V. I., 1961: *Wavefront Propagation in Turbulent Medium*, Dover, New York.

Thompson, L. A., and C. S. Gardner, 1987: Experiments on laser guide stars at Mauna Kea Observatory for adaptive imaging in astronomy. *Nature*, **328**(16), 229–231.

Titterton, P. J., L. E. Mallery, and T. A. Arken, 1971: Lightweight thermosonde system. GTE Sylvania, Inc., Final Report, Contract NAS5-11493.

TMT Science Advisory Committee, 2007: Thirty meter telescope: Detailed science case, TMT.PSC.TEC.07.003.REL01. [Available at http://www.tmt.org/foundation-docs/TMT-DSC-2007-R1.pdf.]

Tokovinin, A., and E. Viard, 2001: Limiting precision of tomographic phase estimation. *J. Opt. Soc. Am. A*, **18**(4), 873–883.

Tokovinin, A., 2002a: Measurement of seeing and the atmospheric time constant by differential scintillations. *Appl. Opt.*, **41**(6), 957–964.

Tokovinin, A., 2002b: From differential image motion to seeing. *Pub. Astron. Soc. Pac.*, **114**(800), 1156–1166.

Tokovinin, A., S. Boumount, and J. Vasquez, 2003: Statistics of turbulence profile at Cerro Tololo. *Mon. Not. R. Astron. Soc.*, **340**(1), 52–58.

Tokovinin, A., V. Kornilov, N. Shatsky, and O. Voziakova, 2003: Restoration of turbulence profile from scintillation indices. *Mon. Not. R. Astron. Soc.*, **343**(3), 891–899.

Tokovinin, A., 2004: Seeing improvement with ground-layer adaptive optics. *Publ. Astron. Soc. Pac.*, **116**(824), 941–951.

Tokovinin, A., J. Vernin, A. Ziad, and M. Chun, 2005: Optical turbulence profiles at Mauna Kea measured by MASS and SCIDAR. *Publ. Astron. Soc. Pac.*, **117**(830), 395–400.

Tokovinin A., 2007: Turbulence profiles from the scintillation of stars, planets, and Moon. *Workshop on Astronomical Site Evaluation*, I. Cruz-Gonzalez, J. Echevarra, and D. Hiriart, Eds., Rev. Mex. Astron. Astrophys. (Conf. Series), 31, 61–70. (Available online at http://www.astroscu.unam.mx/~rmaa/.)

Tokovinin, A., and V. Kornilov, 2007: Accurate seeing measurements with MASS and DIMM. *Mon. Not. R. Astron. Soc.*, **381**(3), 1179–1189.

Tokovinin, A., J. Rajagopal, E. Bustos, and J. Thomas-Osip, 2007: Characterizing ground-layer turbulence with a simple lunar scintillometer. *Proc. Symp. on Seeing*, T. Cherubini and S. Businger, Eds., Kona, Hawaii.

Tokovinin, A., M. Sarazin, and A. Smette, 2007: Testing turbulence model at metric scales with mid-infrared VISIR images at the VLT. *Mon. Not. R. Astron. Soc.*, **378**(2), 701–708.

Tokovinin, A., A. Kellerer, and V. Coudé Du Foresto, 2008: FADE, an instrument to measure the atmospheric coherence time. *Astron. Astrophys.*, **477**(2), 671–680.

Travouillon, T., M. C. B. Ashley, M. G. Burton, J. W. V. Storey, and R. F. Loewenstein, 2003: Atmospheric turbulence at the South Pole and its implications for astronomy. *Astron. Astrophys.*, **400**(3), 1163–1172.

Travouillon, T., 2006: SODAR calibration for turbulence profiling in TMT site testing. *Ground-based and Airborne Telescopes, Proc. SPIE*, L. M. Stepp, Ed., Vol. 6267, 626720.

Travouillon T., S. Els , R. L. Riddle, M. Schöck, and W. Skidmore, 2009: Thirty meter telescope site testing VII: Turbulence coherence time, *Publ. Astron. Soc. Pac.*, **121**, 787–796.

Trinquet, H., and J. Vernin, 2006: A model to forecast seeing and estimate C2N profiles from meteorological data. *Publ. Astron. Soc. Pac.*, **118**(843), 756–764.

Trinquet, H., K. Agabi, J. Vernin, M. Azouit, E. Aristidi, and E. Fossat, 2008: Nighttime optical turbulence vertical structure above Dome C in Antarctica. *Publ. Astron. Soc. Pac.*, **120**(864), 203–211.

Tyson, R. K., 1998: *Principles of Adaptive Optics*. Academic Press, Boston, MA, 345 pp. Tyson, R. K., 2000: *Introduction to Adaptive Optics*. SPIE The International Society for Optical Engineering Press, Bellingham, WA, 130 pp.

Van Zandt, T. E., K. S. Gage, and J. L. Warnock, 1981: An improved model for the calculation of profiles of C2N and ε in the free atmosphere from background profiles of wind, temperature, and humidity. *Proc. 20th Conf. Radar Meteor.*, Boston, MA, Amer. Meteor. Soc., 129.

Van Zandt, T. E., J. L. Green, K. S. Gage, and W. L. Clark, 1978: Radar with a new theoretical model. *Radio Sci.*, **13**, 819.

Vedrenne, N., V. Michau, C. Robert, and J.-M. Conan, 2007: Cn2 profile measurement from Shack-Hartmann data. *Optics Letts.*, **32**(18), 2659–2661.

Vernin, J., and F. Roddier, 1973: Experimental determination of two-dimensional spatiotemporal power spectra stellar light scintillation. evidence for a multiplayer structure of the air turbulence in the upper troposphere. *J. Opt. Soc. Amer.*, **63**(3), 270–273.

Vernin, J., and M. Azouit, 1983: Image processing adapted to the atmospheric speckle. II. Remote sounding of turbulence by means of multidimensional analysis. *J. Optics Paris*, **14**, 131.

Vernin, J., and C. Muñoz-Tuñón, 1992: Optical seeing at La Palma Observatory. 1: General guidelines and preliminary results at the Nordic Optical Telescope. *Astron. Astrophys.*, **257**, 811–816.

Vernin, J., and C. Muñoz-Tuñón, 1994: Optical seeing at LaPalma Observatory. 2: Intensive site testing campaign at the Nordic Optical Telescope. *Astron. Astrophys.*, **284**, 311–318.

Vernin, J., 2002: Mechanism of formation of optical turbulence (Invited Speaker). *Astronomical Site Evaluation in the Visible and Radio Range*, J. Vernin, Z. Benkhaldoun, and C. Muñoz-Tuñón, Eds., Astronomical Soc. of the Pacific Conf. Ser., 266, pp. 2–+.

Vernin, J., H. Trinquet, G. Jumper, E. Murphy, and A. Ratkowski, 2007: OHP02 gravity wave campaign in relation to optical turbulence. *Environ. Fluid Mech.*, 7(5), 371–382.

Véron-Cetty, M. P., and P. Véron, 1996: A catalogue of quasars and active nuclei. *ESO Sci. Rep.*, **17**, 1.

Wang, L., G. Chanan, and M. Schoeck, 2007: Atmospheric turbulence profiling using multiple adaptive optics. *Proc. Symp. on Seeing*, T. Cherubini and S. Businger, Eds., Kona, Hawaii.

Werne, J., and D. C. Fritts, 1999: Stratified shear turbulence: Evolution and statistics. *J. Geophys. Res.*, **26**, 439–442.

Werne, J., T. Lund, B. Pettersson-Reif, P. Sullivan, and D. Fritts, 2005: CAP phase II simulations for the Air Force HEL-JTO Project: Atmospheric turbulence simulations on NAVO's 3000-Processor IBM P4+ and ARL's 2000-processor Intel Xeon EM64T cluster. *Proc. 15th DoD HPC User Group Conf.*, Nashville, TN, 1–22.

West, M., 2005: *A Gentle Rain of Starlight: The Story of Astronomy on Mauna Kea*, Island Heritage Publishers, Hawaii, 108 pp.

Wilson, R.W., 2002: SLODAR: measuring optical turbulence altitude with a Shack–Hartmann wavefront sensor. *Mon. Not. R. Astron. Soc.*, **337**(1), 103–108.

Winker, D., 1991: Effect of a finite outer scale on the Zernike decomposition of atmospheric optical turbulence. *J. Opt. Soc. Am. A*, **8**(10), 1568–1573.

Wyngaard, J. C., and O. R. Coté, 1971a: The budget of turbulent kinetic energy and temperature variance in the atmospheric surface layer. *J. Atm. Sci.*, **28**, 190–201.

Wyngaard, J. C., Y. Izumi, and S. A. Collins, 1971b: Behaviour of the refractive index structure parameter near the ground. *J. Opt. Soc. Amer.*, **61**(12), 1646–1650.

Yaglom, A. M., 1949: On the local structure of a temperature field in a turbulent flow (in Russian). *Dokl Adad Nauk SSSR*, **69**(6), 743–749.

Ziad, A., R. Conan, A. Tokovinin, F. Martin, and J. Borgnino, 2000: From the grating scale monitor to the generalized seeing monitor. *Appl. Opt.*, **39**(30), 5415–5425.

Ziad, A., J. Borgnino, F. Martin, J. Maire, and D. Mourard, 2004a: Toward the monitoring of atmospheric turbulence model. *Astron. Astrophys.*, **414**(3), L33-L36.

Ziad, A., M. Scheock, G.A. Chanan, M. Troy, R. Dekany, B. F. Lane, J. Borgnino, and F. Martin, 2004: Comparison of measurements of the outer scale of turbulence by three different techniques. *Appl. Opt.*, **43**(11), 2316–2324.

Ziad, A., E. Aristidi , A. Agabi, J. Borgnino, F. Martin, and E. Fossat, 2008: First statistics of the turbulence outer scale at Dome C. *Astron. Astrophys.*, **491**(3), 917–921.